Animals and People
Sharing the World

ANIMALS AND PEOPLE SHARING THE WORLD

Edited by
Andrew N. Rowan

Published for Tufts University
by University Press of New England
Hanover and London

© 1988 by the Trustees of Tufts University

All rights reserved. Except for brief quotations in critical articles or reviews, this book, or parts thereof, must not be reproduced in any form without permission in writing from the publisher. For further information, contact University Press of New England, Hanover, New Hampshire 03755.

Printed in the United States of America.

5   4   3   2   1

Library of Congress Cataloging in Publication Data

Animals and people sharing the world.

     Bibliography: p.
     1. Pets–Social aspects. 2. Pet owners.
3. Human–animal relationships. 4. Domestic animals–
Social aspects. 5. Animals, Treatment of. I. Rowan,
Andrew N.
SF411.4.A56 1988 636 88-40114
ISBN 0-87451-449-5
ISBN 0-87451-465-7 (pbk.)

Contents

family, a sensitive professional, a man of compassion, and a worker to improve the quality of life for all."

This volume, consisting of contributions from selected speakers at the Delta Society international conference in Boston, is dedicated to the memory of Michael McCulloch. We who knew him or knew of him must now realize the vision that he will never see fulfilled.

Andrew N. Rowan

Boston, Massachusetts
May 1988

Preface

The Delta Society is the brainchild of a small group of sensitive professionals who recognize the importance of animals as companions. Without their insight and hard work, pets might be absent from the lives of many people, particularly senior citizens, criminals, and the handicapped.

One of Delta's founders, Michael McCulloch, was especially committed to the promotion of research, legislation, and programs essential to preserve the bond between people and animals. A clinical instructor in the Department of Psychiatry, University of Oregon Health Sciences Center, and associate clinical professor of psychiatry in veterinary medicine at Oregon State University, McCulloch was himself a pioneer in the use of pets for therapy. He lectured nationwide on the importance of pets in our lives, emphasizing intelligent stewardship: "If pet therapy offers hope for relief of human suffering, then it is our professional obligation to explore every available avenue for its use."

On June 16, 1985, Michael McCulloch was in h prime, vice president of the Delta Society, workir energetically to raise funds to provide a solid financ base for Delta's future growth. By the end of the day, was dead, shot by one of his psychiatric patients in office.

Michael McCulloch's death diminishes us al colleague and winner of the first Michael McCu Award sums up the loss in his acceptance speech have lost a husband, a father, a member of an ext

Biographical Information on the Contributors

Alan M. Beck, ScD

Dr. Beck received his Baccalaureate from Brooklyn College, Masters degree from California State University at Los Angeles and Doctorate in Animal Ecology from The John Hopkins University School of Hygiene and Publiic Health. He directed the animal programs for the New York City Department of Health for five years. Presently, he is Director of the *Center for the Interaction of Animals and Society* at the University of Pennsylvania's School of Veterinary Medicine. Dr. Beck is a founding Board Member of the Delta Society.

His book, *The Ecology of Stray Dogs*, is considered a classic in the field of urban ecology. Together with Dr. Aaron Katcher, he edited the book *New Perspectives on Our Lives with Companion Animals* and authored the popular book, *Between Pets and People: The Importance of Animal Companionship*.

Gordon M. Burghardt, PhD

Dr. Burghardt is a Professor of Psychology and Zoology at the University of Tennessee in Knoxville. He is also Director of the Ethnology section of the Graduate Program in the Life Sciences at the University of Tennessee. He received his PhD in psychology from the University of Chicago and is past president of the Animal Behavior Society. He has published extensively on a wide range of topics, including the development of behavior, the behavior of reptiles, and human-animal interactions.

Marianna Burt, MA

Ms. Burt, currently a law student at the University of North Carolina, completed her PhD coursework at the University of Chicago and has taught humanities and art history at several universities. She has been a Delta Society member since 1980 and joined the Delta Society Board of Consultants in 1983. That year she was co-author of the Farnsworth House study, one of the first projects to investigate pet ownership among elderly/ handicapped persons living in independent settings. Active in humane work for more than twenty years, she currently administers the PATH program (People and Animals Together for Health) and is a member of the Board of Directors of the North Carolina Federation of Humane Societies.

Bruce Fogle, DVM, MRCVS

Dr. Fogle is a private practitioner and owner of the Portman Veterinary Clinic in central London. He received his DMV from the Ontario Veterinary College in Guelph and went to London on a postgraduate travel scholarship where he established his own private practice. He is a founder member of the Society for Companion Animal Studies and, in 1980, organized an international symposium in London on the human-animal bond. He is editor of *Interrelations between People and Pets* and author of *Pets and Their People*.

Harold A. Herzog, PhD

Dr. Herzog is Associate Professor of Psychology at Western Carolina University in Cullowhee, North Carolina. He received his PhD in psychology from the University of Tennessee. He has long-standing interest in the human-animal bond and has studied a wide variety of human-animal interactions, including cock-fighting, slaughtering, the attitudes of veterinary students, and the approach of the tabloid newspapers to animals.

Aaron H. Katcher, MD

Dr. Katcher is associate professor of psychiatry at the University of Pennsylvania and associate director of the Center for the Interaction of Animals and Society. His long-standing interest in the influence of social conditions on emotional states and health led him to study the health consequences of pet ownership. His studies have used experimental, ethological, and psychophysiological techniques to describe the interactions between people and pets. Most recently he has enlarged the context of his studies to examine the influence of a much wider range of interactions with the living components of the human environment.

Stephen R. Kellert, PhD

Dr. Kellert is currently a Professor at the Yale School of Forestry and Environmental Studies, where he conducted numerous studies on public attitudes to wildlife. He is Chairman of the Connecticut Nonharvest Wildlife Board, and past chairman and founder of the Human Dimensions in Wildlife Study Group. In 1983, he received a special conservation award from the National Wildlife Federation. In 1986, he was awarded a Fulbright Scholarship to study attitudes to wildlife in Japan and, in 1987, he received an award for the best publication from the International Foundation for Environmental Conservation.

Elizabeth A. Lawrence, VMD, PhD

Dr. Lawrence is an Associate Professor in the Department of Environmental Studies at Tufts University School of Veterinary Medicine. She received her BA from Mount Holyoke College, MA, her VMP from the University of Pennsylvania School of Veterinary

Medicine, and her PhD in cultural anthropology from Brown University, RI. She has wide-ranging research interests including cross-cultural aspects of horse-human interactions, native American views of animals and nature, the role of animals as symbols, and the history of veterinary medicine. She is the recipient of two national awards for manuscripts dealing with human-animal interactions and is the author of three books: *Rodeo: An Anthropologist Looks at the Wild and the Tame, Hoofbeats and Society: Studies of Human-Horse Interactions,* and *His Very Silence Speaks.*

Harriet Ritvo, PhD

Dr. Ritvo is an Associate Professor in the Humanities Department of the Massachusetts Institute of Technology. She is the author of *The Animal Estate: The English and Other Creatures in the Victorian Age.* Her articles about Victorian England and attitudes toward animals have appeared in a variety of periodicals, including *Victorian Studies, Bioscience, Comparative Studies in Society and History,* and *Science, Technology, and Human Values.*

Andrew N. Rowan, DPhil

Dr. Rowan is Associate Professor in the Department of Environmental Studies at Tufts University School of Veterinary Medicine and Director of the Tufts Center for Animals and Public Policy. He received his DPhil in biochemistry from Oxford University and subsequently worked for several animal societies on the issue of animal research and alternatives. He is the author of *Of Mice Models and Men* and the editor of *Anthrozoös,* the official journal of the Delta Society that covers human-animal-environment interactions. He is on the Board of Directors of the Delta Society as well as of several other non-profit groups.

James A. Serpell, PhD

Dr. Serpell is Director of the Cambridge Companion Animal Research Group based in the Sub-Department of Animal Behaviour at Cambridge University. He received a zoology degree from University College, London, and his PhD in animal behavior from the University of Liverpool. He has been a leading figure in research on the human-animal bond and is author of the comprehensive text, *In the Company of Animals.*

Animals and People
Sharing the World

Introduction: The Power
of the Animal Symbol
and Its Implications

Andrew N. Rowan

Anyone who campaigns for the improved welfare of animals sooner or later faces the accusation that worrying about animals is a waste of time when so much human suffering needs to be addressed. Although this accusation is not easy to refute, most animal activists would stress the connections between harm to animals and harm to humans and the environment, arguing that animal welfare is part of a world crisis that cannot be broken down according to problems of human or animal welfare, urban or rural pollution, nuclear or conventional arms. Anybody who works to improve the well-being of humans, animals, or the environment contributes to the advancement of planetary health and human civilization. As Gandhi once remarked, "the greatness of a nation and its moral progress can be judged by the way its animals are treated" (Gandhi, 1959).

For the most part, this and similar arguments leave the critic without an effective rejoinder; yet this did not prevent me from constantly reexamining my role in animal welfare activities—should I campaign instead to eliminate world hunger or to provide clean, potable water for the majority of the world's inhabitants who have none? A few years ago, however, I was introduced to Leach's remarkable analysis (1964) of animal categories and verbal abuse, an article that provided me with new insight into the importance of animals to human existence and renewed confidence that my work on behalf of animal welfare was worthwhile and could be justified in the broader context of a diverse, compassionate, and sustainable human existence.

Animal Terms and Terms of Abuse

The arguments in Leach's rich, complex paper plumb several levels of meaning. They are thus very difficult to summarize, especially for a biochemist—like myself—with only a very superficial understanding of anthropological literature and concepts. Nevertheless, just as one does not have to understand the nuances of astrophysics to comprehend the gist of the Big Bang theory, so one does not have to be a trained anthropologist to benefit from Leach's arguments.

Leach (1964) notes that the language of obscenity (regardless of the language studied) falls into three broad categories:

a. "dirty" words (referring to sex or excretion);
b. blasphemy and profanity; and
c. terms by which a human is likened to an animal.

One does not have to be a psychiatrist or a psychologist to see that terms dealing with sex and excretion can have linguistic potency. Similarly, even in the modern secular age, it is understandable that blasphemy and profanity arouse or release emotions. However, animal categories of verbal abuse seem less easily accounted for. In Leach's words (1964), "when an animal name is used . . . as an imprecation, it indicates that the name itself is credited with potency. It clearly signifies that the animal category is in some way taboo and sacred. Thus, for an anthropologist, animal abuse is part of a wide field of study which includes sacrifice and totemism." In the simplest (and most provocative) terms, animal symbols appear to be as important to humans as God and sex. For those of us who study human-animal interactions or campaign on behalf of the welfare of animals, it is reassuring to be involved in a phenomenon of some social importance, rather than a side issue of marginal relevance.

Unfortunately, it is characteristic of academic discourse that any paper that agrees with one's own intuitive ideas (or innate wisdom!) eventually will be attacked by somebody with a contrary argument. Thus, in 1976, anthropologist John Halverson published a stinging critique of Leach's paper deploring the loose and varied use of the term "taboo," identifying a host of errors in Leach's etymological scholarship, and arguing that the negative connotations associated with animal terms simply might reflect the basic human-versus-animal conflict, not some deeper taboo related to human difficulties with categorizing domestic animals as "us" or "not-us." Halverson's arguments are detailed and convincing, and yet he may go too far in the other direction, reducing the place of animals in human life to the merely mundane and eliminating animals and animal terms as symbols of varying degrees of potency.

Killing of Animals, Domestic and Wild

Humankind has always found it difficult to assign an appropriate status to animals within its changing worldviews. Hunter-gatherer societies indulge in elaborate propitiation ceremonies to appease the spirits or spirit-guardians of the animals they have killed. Modern societies appear—superficially, at least—to have overcome their guilt over killing animals; yet a closer examination suggests that this surface may conceal a deep-seated and unexamined guilt (the common use of the term "sacrifice" in animal research reflects an aspect of our concern over animal death). In kosher slaughter, the actual killing is performed by a holy man, the *shechita*, because he has the grace and strength of character to bear the associated guilt. In Tibet, butchers are considered outcasts. Recreational hunters of the present day, especially those such as Aldo Leopold who have achieved a significant communion with nature, enjoy the hunt and the kill yet agonize over the paradox that their bonds with nature seem strongest when they kill one of its denizens (Elder, 1986).

Elder's article (1986) is a superb introduction to the contradictions within a man like Leopold, who extolls hunting as a way of life (of being wild himself) but also affirms the personhood and the rights of other living creatures. Leopold is not blind to this conflict and attempts time and again to come to terms with the ethical and evolutionary challenge. Early in his life, he killed a wolf (believing that the more wolves were killed, the more deer would prosper) but, after watching the green fire dim and die in the eyes of the mortally wounded creature, he began to question his approach. His quest led him eventually to the "Land" ethic, but he still hunts.

It is fitting that Leopold's consciousness was stirred by a dying wolf, since wolves and bears seem to carry special meaning and symbolism for humans. For example, in Norway, the traditional wolf range now carries millions of sheep, millions of people, and five to ten wolves. It would seem that so few wolves should not disturb people, but the recent sighting of a wolf in southern Norway sparked a frenzied wolf hunt and a lucrative tourist trade in wolf artifacts that did not end until a wolf was shot dead, doused with champagne, and dragged off first to a school, then to an old people's home, and finally to the steps of the national parliament. Even today, with modern weapons, sophisticated transport, and a highly educated population, one solitary wolf is still capable of inciting mass hysteria.

Human Language and Human Uniqueness

The debate over ape language also reveals much about our insecurities over the place of animals in our anthropocentric worldview. Elizabeth Anscombe, the Cambridge philosopher, was once asked what she thought people would do if it could be shown that apes could speak. Reportedly, she responded, "Simple, they'll up the ante" by redefining language so that it would remain uniquely human. The bitter, partisan nature of the ape-language debate among otherwise objective scholars (Fowler, 1980) supports the notion that we cannot deal rationally and objectively with the possibility of a breach in humankind's last defense of its unique position—language. Eugene Linden's recent book, *Silent Partners* (1986), describes some of the prejudices and passions underlying ape-language research. Linden deplores that so much time is spent trying to support or refute arguments and data about linguistic sophistication that we are ignoring the new ape-human communication systems' potential to provide fascinating insights into the worldview of the Pongidae. For example, when Viki Hayes, a chimp raised in a human household, was asked to sort human and animal photographs into categories, she placed the pet dogs' pictures on the pile of animal photographs but her own photograph with the human snapshots (cited in Goodall, 1986).

Modern Pets/Companion Animals

A little closer to the average suburban home, one finds many other examples of the ambiguous role played by animals in our lives—namely, our pets. Most of us name our pets, and, as anthropologists recognize, naming is not a random process. A minority of pets (22% of dogs and 10% of cats) are given common human first names; the rest are given non-human names (59% of dogs and 78% of cats) or human names that are foreign or out of fashion (Rowan, 1987). The significance of this may be heightened by considering another culture, the Beng people of West Africa, who also keep dogs. These animals are rarely shown any affection and must fend for themselves. If they become sick, however, they are cared for and nursed back to health. Beng dogs are also given names, but these are nonsense words or are words from foreign languages that express negative or fatalistic concepts (Gottlieb, 1986).

The dog occupies an ambiguous position in Beng mythology: It introduced death to humanity but, in another myth, saves humanity while betraying its fellow animals. The naming of dogs, but only with non-Beng words, reflects this ambivalent role. Similarly, the names (and nicknames) given to our pets may reflect our own underlying estimations of our animals. The choice of a human name might, for example, indicate that an animal has a relatively high status.

Another feature of American and European pet-keeping is the disparate attitudes toward cats and dogs. While this phenomenon is well-known and commonplace, there have been few, if any, studies of the reasons underlying the differences. Selby and Rhoades (1981) report that only 4% of their sample disliked dogs, but 28% disliked cats (see Table 1).

It is probable that the behavioral patterns of dogs and cats are responsible. Once they have accepted a person as pack leader or member, dogs respond with exaggerated enthusiasm and affection. They are submissive and cannot hold a person's gaze for any length of time. By contrast, cat behavior appears far more deliberate and selective, and cats can stare unblinking at a human in a quite disconcerting way. Our attitudes to dogs and cats, and also to specific breeds of dogs and cats, probably reflect to a large degree our own personalities and insecurities and the type of image we would like to project upon the world. The need for reinforcement of a macho ego is all too apparent in the young male who

**Table 1. Attitudes to Dogs and Cats
of a Sample of 910 Owners and Non-Owners.**

Attitude Statement: "I dislike dogs/cats"	Dogs (%)	Cats (%)
Strongly agree	1.5	11.9
Agree	2.5	15.9
Neutral	7.1	15.3
Disagree	33.2	31.4
Strongly Disagree	53.6	22.0
No opinion	2.1	3.5

obtains a pit bull terrier and a studded leather collar and chain to go with it.

Animal companions, however, are more than just smelly creatures to be fed, walked, and cleaned up after. Companion animals offer a safe outlet for human needs for physical contact with another warm being and may satisfy some human needs for intimacy. According to a recent story in the *New York Times* Science Section (November 3, 1987), people indicate that intimacy (in friendship) is increasingly important to their sense of satisfaction with life. Clearly, pets provide one outlet for expression of physical intimacy, but nobody has explored the correlation between pet-keeping and an individual's need for intimacy. Regarding touching behavior, a recent study reported in *Science* (Barnes, 1988) finds that neonates experience a biochemical response to their mother's touch that promotes growth. This response could be bidirectional, with both adult and neonate benefiting from physical contact.

Delta Society Plenary Papers

The ensuing papers, presented at the Delta Society's International Conference on People, Animals and the Environment in Boston in 1986, all examine, explicitly or implicitly, the symbolic role of animals in modern life.

Ritvo explores the history of pet-keeping in the industrialized world, especially our fascination with creating new and ever-more-exotic breeds. Serpell studies the long-range history of pet-keeping and illustrates some peculiarities of the practice, as well as its enduring and cross-cultural nature. Katcher and Beck, in a lyrical essay, speculate on the importance of nurturing living beings in promoting the maturation of children and the maintenance of human health. Herzog and Burghardt review modern Western attitudes to animals and, in particular, the "popular" view as exemplified by the more lurid examples of the yellow press. Lawrence brings insight, compassion, and broad-ranging scholarship to her analysis of the horse-human interaction. Burt surveys the representation of animals or animal symbols in human art and discusses its importance in human existence. Kellert examines human attitudes (in North America) to wildlife and nature. Finally, Fogle rounds off the collection with a concluding statement.

The one obvious gap in this volume is the absence of a general review of animal-child interactions. In fact, considering the ubiquity of pets in modern industrial societies, it is surprising how little literature exists on the effect of pets on the mental life and development of children. Piaget (1929) noted that all children go through a phase during which they ascribe human traits to animals (no doubt accounting in part for Disney's popularity), and Freud (1953) observed that children have no scruples about ranking individual animals as their equals (indeed, adults may be much more puzzling to young children than the animals in the home). This does not, of course, mean that animals are always liked—they evoke a range of feelings from awe and love to hate and fear, and it has been reported that animals are among the most common phobias of young children (Bauer, 1976).

Despite Kellert's review, the human-environment interaction is also underrepresented here, although there is an extensive literature on this topic, detailed in a recent bibliography (Kellert and Berry, 1985).

Conclusion

Ever since human beings appeared on the African savannah, other animals have been an essential part of our history, culture, and existence. It is hardly surprising that we have strong attitudes about our relationships with them—as strong, perhaps, as our attitudes to such topics as sex, religion, and politics. The old but often true cliché that politicians receive more mail from constituents on animal issues than on any other subject, including abortion and the Vietnam war, hints at the strength of our feelings. When army nerve-gas tests on dogs were publicized in 1974, the Pentagon received 30,000 protest letters—more than when General MacArthur was fired (Holden, 1974).

Animals have been, and are, viewed in many different ways by different cultures and peoples. As mentioned, hunter-gatherers make sacrifices to appease the god of the animals they kill for food. The ancient Greeks developed a fairly detailed moral classification for animals long before Aristotle introduced his scale of being. Foxes were perceived as crafty, asses were slow, and dogs were generally despised.

Interestingly, most ancient peoples reviled the dog; our current preoccupation with and concern for the canine is a complete reversal of ancestral attitudes. The dog is the American "sacred cow" (Beck, 1974), and every president is photographed posing with a dog. But dogs are generally regarded as vermin in other parts of the world—even today. The municipal authorities in Peking have banned dogs—except for police dogs and those bred for meat—because they consider them a danger to health and society. In Russia, there has been a recent, sharp increase in anti-dog letters to the press (IJSAP, 1983). There are calls to destroy all dogs except service animals. Ownership of dogs is perceived as anti-proletarian, because tons of food, which could nourish humans, are consumed by animals instead. In addition, dog-control measures are virtually nonexistent, and there are reports that hundreds of thousands of people are annually attacked and bitten by roaming dogs. In modern Western civilization, most of us anthropomorphize animals in one way or another, and we are frequently irrational in the way we regard and interact with them. In some cases, attitudes to animals are influenced directly by religious beliefs. For example, those who believe in the transmigration of souls between humans and animals tend to treat animals differently from those who do not. Consider the ancient Greek joke in which Pythagoras reportedly stopped a man from beating a dog because he (Pythagoras) recognized, in the dog's yelps, the cries of a dead friend. The joke has obviously suffered in translation, but the contempt felt for dogs in those days presumably contributed to the satire.

Our attitudes to animals, our use of animal symbols, and the significance attached to such symbols (the Exxon tiger in the tank, the eagle of the United States) have been subjected to relatively little careful review and study. Animal symbols are usually taken for granted, and even respected scholars are likely to dismiss phenomena relating to animals as unimportant. It is hoped that this volume, building on the proceedings of earlier human-animal bond meetings (Anderson, 1975; Katcher and Beck, 1983; Anderson et al., 1984), will add to the growing momentum of scholarship in this field to enrich our understanding of ourselves and the animals with which we share the world and to develop human compassion and civilization.

References

Anderson, R. K., Hart, B. L., and Hart, L. A., eds. 1984. *The Pet Connection.* Minneapolis: Center to Study Human–Animal Relationships and Environments, University of Minnesota.

Anderson, R. S., ed. 1975. *Pet Animals and Society.* London: Bailliere Tindall.

Barnes, D. M. 1988. Need for Mother's Touch is Brain-Based. *Science 239*:142.

Bauer, D. H. 1976. An Exploratory Study of Developmental Changes in Children's Fears. *Journal of Child Psychology and Psychiatry 17*:69–74.

Beck, A. M. 1974. The Dog: America's Sacred Cow? *Nation's Cities 12(2)*:29–31, 34–5.

Elder, J. 1986. Hunting in Sand County. *Orion Nature Quarterly 5(4)*:46–53.

Fowler, S. 1980. The Clever Hans Phenomenon Conference. *International Journal for the Study of Animal Problems 1*:355–59.

Freud, S. [1913] 1953. *Totem and Taboo*, standard ed., *13*:1–161. London: Hogarth Press.

Gandhi, M. 1959. The Moral Basis of Vegetarianism. Ahmedabad, India: Novajivan Publishing.

Goodall, J. 1986. *The Chimpanzees of Gombe: Patterns of Behavior*, 571–86. Cambridge: Harvard University Press.

Gottlieb, A. 1986. Dog: Ally or Traitor? *American Ethnologist 13*:477–88.

Halverson, J. 1976. Animal Categories and Terms of Abuse. *Man 11*:505–16.

Holden, C. 1974. Beagles: Army Under Attack for Research at Edgewood. *Science 185*:130–31.

IJSAP. 1983. New assaults on dogs in the USSR. *International Journal for the Study of Animal Problems 4*:95.

Katcher, A. H., and Beck, A. M., eds. 1983. *New Perspectives on Our Lives with Companion Animals.* Philadelphia: University of Pennsylvania Press.

Kellert, S. R., and Berry, J. K. 1985. *A Bibliography of Human/Animal Relations.* Lanham, MD: University Press of America.

Leach, E. R. 1964. *Anthropological Aspects of Language: Animal Categories and Verbal Abuse.* In *New Directions in the Study of*

Language, 23–64, ed. E. Lennenberg. Cambridge, Mass: MIT Press.

Linden, E. 1986. *Silent Partners*. New York: Times Books.

Piaget, J. 1929. *The Child's Conception of the World*. New York: Harcourt Brace.

Rowan, A. 1987. Editorial. *Anthrozoös* 1:63–4.

Selby, L. A., and Rhoades, J. D. 1981. Attitudes of the Public Towards Dogs and Cats as Companion Animals. *Journal of Small Animal Practice* 22:129–37.

The Emergence of Modern Pet-Keeping

Harriet Ritvo

Today, pets figure so prominently in our lives that it is difficult to imagine pet-keeping as the subject of satire and ridicule. And yet, two hundred years ago, a prominent Englishman who wished to make provision for his two pet dogs in his will, hid the bequest in a secret codicil to avoid hostile public opinion. Harriet Ritvo argues that pet-keeping in its modern form is, like animal protection societies, a Victorian invention, the motives for which were rather mixed.

It is probably no accident that pet-keeping and concern for animals developed simultaneously. The beginning of widespread pet ownership also coincided with increasing human mastery over the caprices of the natural world. Wilderness, which had been castigated as ugly, increasingly became the subject of aesthetic appreciation. The safe, captive, and loyal pet was an obedient and subservient symbol for the appropriate relationship between humankind and the natural world. Control of the breeding of pets allowed owners to shape their animals' form and behavior. This spurred the development of the dog fancy in the mid-nineteenth century. Interestingly, pet-keeping by the lower classes was condemned as an unwarranted indulgence, ostensibly because it led to the neglect of their children. However, the real concern may have been that pet-keeping was considered a suitable prerogative only for those who exercised similar control over their fellow human beings. EDITOR

Nowadays few people feel the need to conceal their affection for their pets. On the contrary, many contemporary English and American households unashamedly include dogs and cats, and the owners of particularly elegant or surprising pets are inclined to flaunt them as symbols of discriminating taste or conspicuous consumption. The economic importance and social respectability of pet-keeping is further attested by the flourishing institutions—from animal hospitals to breed clubs to the pet food industry—that service pets and their owners. A great deal of our society's time, energy, and money is now devoted to satisfying pets' and pet owners' needs, physical and otherwise (Kellert, 1983; Serpell, 1986).

When thinking about the history of household animals, we are often tempted simply to project into the past the conditions that we daily observe in the present. Pets figure so prominently in contemporary American and European life that it is difficult to imagine earlier versions of our societies in which they did not enjoy similar appreciation. Yet if we look back as little as two centuries, the position occupied by domestic dogs and cats begins to seem unfamiliar. One measure of this difference is the paucity of evidence about the relationships between humans and their animal companions. Because household animals were not considered objects of serious inquiry or widespread appeal, few authors and publishers were inclined to spend time and ink on them. These intellectual and economic disincentives were reinforced by more direct social coercion. Individuals who expressed untoward interest in or affection for pet animals might provoke disapproval, contempt, or suspicion.

An example of the kind of source in which information about eighteenth-century pets is apt to be embedded is the *History of the Ancient and Honorable Tuesday Club,* an extremely long satirical account of a club for upper-class men, which met regularly in

Annapolis, Maryland, during the 1740s and 1750s. The author, a Scottish emigrant named Alexander Hamilton, apparently wrote for a restricted audience, since his narrative is only now being prepared for publication. Although the *History* has certain literary pretensions— for example, it is written in a high mock-heroic mode— it is primarily of interest today as a storehouse of social history. As he wryly chronicled the annals of the Tuesday Club, Hamilton touched on a variety of everyday topics, including attitudes toward pets and pet-keeping. For example, one of the leading figures in the Tuesday Club was fond of cats, and the satirist made this predilection the emblem of his target's luxurious and depraved character. The following is an excerpt from what is intended as a biting portrait:

> *This celebrated Gentleman, Judging his own Species, unworthy to make constant companions and Intimates of, Chose a Society of Cats for his friends, fellows and playmates, both at bed and board, and so far did his extraordinary charity and benevolence extend to those Cats, that he would deign to converse with them in the most familiar manner, giving some of them a christian like education ... it is said, that he once buried a favorite Cat, with great form and ceremony, like a christian.... Some may think it very Strange, that ... a Gentleman born and bred in a christian land, should pay so much deference and respect, to these brute creatures.... In fine, had he been a persian Dervis, who understood thorro-ly the Language of brutes, there might be some plausible reason for his amusing himself in this manner; but as he is an old Englishman, and a protestant, and a Christian of the Church of England, as by Law established, there is no other way ... for accounting for this odd humor, but by ascribing it to mere whim and fancy.... (Hamilton, in press).*

This selection, which is extracted from a much longer diatribe in the same vein, suggests that pet-keeping was not only unusual in the American colonies on the eve of Revolution, but was also an easy target for criticism and ridicule. Other contemporary sources corroborate the low esteem in which pet ownership was held. In England at about the same period Humphrey Morice, a gentleman of some distinction who had served as a Privy Councillor, wished to provide for the maintenance of his aged dogs after he died. Because he feared public opinion, however, he was reluctant to include this bequest in the main body of his will, hiding it instead in a secret codicil, which he cast in the form of a letter to a friend (Harwood, 1928). More generally, the metaphorical or symbolic or connotative penumbra surrounding the dog, which was then as now the archetypal pet species, although not the only one, indicated the low regard in which both animals and their owners might be held. In the eighteenth century the main characteristics attributed to the dog were not the loyalty and affection that predominate in current iconography. Instead, as many of William Hogarth's satiric paintings and engravings demonstrated, the dog was more likely to represent bestiality, vulgarity, and subversion. Thomas Bewick, the author and illustrator of popular works of natural history, wrote that, although the dog was loyal to his master, "to his own species he is ill-behaved, selfish, cruel, and unjust." Earlier still, during the sixteenth and seventeenth centuries, the contrast with modern understandings of the dog's moral nature was still more striking. Shakespeare, for example, seldom referred to dogs except to express his distaste for them and for the people with whom he associated them (Bewick, 1975; Empson, 1951; Paulson, 1979).

It is necessary to emphasize that English and American pets have only recently gained widespread acceptance because it is so easy to assume the reverse. Americans routinely think of their own culture, and especially the part of it that derives from Great Britain, as one that embodies a unique, intrinsic, and immemorial fondness for animals. English self-stereotypes echo this point of view. Thus when a publican in the western part of England was recently prosecuted under the anti-cruelty laws for advertising hedgehog-flavored potato chips, the media on both sides of the Atlantic reported it as typical, lovable English dottiness. And there is more solid, less anecdotal evidence for this understanding of Anglo-American attitudes. For example, the institutionalized animal protection movement originated in Great Britain with the foundation of the (not yet Royal) Society for the Prevention of Cruelty to Animals in 1824; the Cruelty to Animals Act of 1876 pioneered state regulation of animal experimentation (French, 1975; Turner, 1980). But this heightened sensitivity to animal suffering did not develop in a vacuum; the activists who crusaded on behalf of their fellow creatures were inspired not by benevolent abstractions but by repeated observation of violent physical abuse. However early nineteenth-century Britons may have stacked up in relation to the citizens of other nations, most of them were more notable for their indifference to animal pain, or even their enjoyment of it when it came in the guise of such sports as dog fighting, badger baiting, or rat killing, than for their abhorrence of cruelty to their fellow creatures (Ritvo, 1987).

Thus the tradition of extravagant British concern for animals, like many other apparently ancient usages, turns out to have been a Victorian creation (Hobsbawm and Ranger, 1983). Although at the end of the nineteenth century a humanitarian crusader named W. J. Stillman could celebrate the strength, "especially in Anglo-Saxon countries," of "this sentiment of tenderness for...the sentient lower creatures," English self-characterizations of only slightly earlier offered quite a different impression. In the 1830s, for example, the naturalist Edward Jesse wrote that "of all the nations of Europe, our own countrymen are, perhaps, the least inclined to treat the brute creation with tenderness," and in 1868 Queen Victoria, who was herself both an enthusiastic partisan of animals and a devoted pet owner, complained that "the English are inclined to be more

cruel to animals than some other civilized nations are."
Even after the Victorian period, some human victims of
oppression, especially women, but also members of the
lower classes, found an easy metaphor for their suffer-
ings in those routinely inflicted on laboratory animals
(Jesse, 1835; Stillman, 1899; Hibbert, 1984; Lansbury,
1985).

In the unpropitious emotional or moral climate of
pre-Victorian Britain (although most of the following re-
fers specifically to Britain, it also applies to the history
of pet-keeping in America, where, at least through the
nineteenth century, developments tended after a slight
time lag to parallel those across the ocean), it is perhaps
not surprising that few people kept pets. That is, although domesticated animals, including dogs and cats,
had been part of many households from Saxon and
Celtic times, almost all of these creatures were kept because they performed some useful function, not simply
because they were affectionate or ornamental. A six-
teenth-century listing of English dog breeds divided the
animals by function: for example, "Fynder," "Stealer,"
"Turnspit," and "Dancer" (Caius, undated). Even dogs
that kept relatively elevated human company, such as
those that appeared with their masters in the formal
portraits of the seventeenth and eighteenth centuries—
portraits are often adduced as evidence of the antiquity
of pet-keeping—were not pets, but rather hunting dogs
like setters and spaniels or, more rarely, coursing dogs
like greyhounds.

Nevertheless, pet-keeping did have a long history
among certain kinds of people. A medieval example,
admittedly fictional but representative of a trend among
her small group of peers, was the privileged prioress of
Chaucer's fourteenth-century *Canterbury Tales*, who
traveled with "smale houndes" that she fed prodigally
"with rosted flessh, or milk." In Chaucer's view, and in
that of the ecclesiastical authorities of his day, who tried
to suppress pet ownership among monks and nuns

from affluent backgrounds, such behavior was inappropriate. Despite this official censure, however, pet-keeping among the very privileged did not disappear, and, a century and a half later, the secular ladies attending the court of Henry VIII were allowed to bring their dogs (Thomas, 1983). Beginning in the seventeenth century the owners of estates occasionally erected monuments in memory of deceased animals, although it is usually difficult to tell from the surviving inscriptions whether the commemorated animal was a genuine pet or merely a loyal servant (Lambton, 1985). And during the latter part of the seventeenth century, King Charles II was renowned—or perhaps notorious—for doting on lapdogs, and his fondness has been preserved in the name of one of the major varieties of toy spaniels. When his pets were stolen, as seems to have happened with some frequency, he was inconsolable; once he went so far as to advertise in a newspaper for a favorite's return. (He may have been one of the first to avail himself of this method of repairing his loss, since newspapers were a recent innovation at that time.) Charles' brother and successor, James II, also enjoyed pet dogs, as did his successors, William and Mary. During their joint reign the pug, which was, like William, a native of the Netherlands, became established as the preferred lapdog of the English aristocracy (Ritchie, 1981).

What these early pet owners had in common was privileged status in terms of both money and rank. This meant, on the crudest material level, that they could afford to maintain animals that did not earn their keep. It also gave them sufficient independence to ignore any criticism or derision that might be directed their way. And on a deeper level, they may have enjoyed a metaphorical security—a feeling of supremacy over nature—that was as unusual as was their exalted social position. For animals, even heavily domesticated pet animals, have always symbolized the natural world, and incorporating one into the intimate family circle would have presupposed an attitude of trust and confidence that few ordinary English citizens of the sixteenth, seventeenth, and even much of the eighteenth centuries were able to muster. Pet owners probably saw the non-human world as a less threatening and more comfortable place than did most of their contemporaries, who understood their relationship with the forces of nature primarily as a struggle for survival. That is why pet-keeping did not become a widely exercised prerogative until this struggle had been sufficiently mediated or attenuated by scientific, technological, and economic developments. Only at that point could ordinary people interpret the adoption of a representative of the elements (however tame and accommodating) as reassuring evidence of human power, rather than as a troublesome reminder of human vulnerability (Passmore, 1975; Shell, 1986; Thomas, 1983).

Thus it is not surprising that widespread pet ownership among members of the middle classes can be dated from the late eighteenth and early nineteenth centuries. This period saw a series of radical changes in the general relationship of human beings (at least European human beings) to the natural world. Whereas at the beginning of the eighteenth century natural forces had been perceived as largely out of human control, by the end of the century science and engineering had begun to make much of nature more manageable. Advances in natural history, and especially in taxonomy, signaled an increase in human intellectual mastery. On a more pragmatic level, progress in such fields as animal husbandry, veterinary science, and weapons technology made those who had to deal with animals less vulnerable to natural caprice. These technical developments were paralleled in the political sphere by the increase of English influence in those areas of the world—Asia, Africa, and North America—where nature was perceived to be wildest. Once it had become the subject of domination rather than a constantly menacing antagonist, nature could be viewed with affection and even, as the scales tipped more to the human side, with nostalgia. This shift had consequences throughout western culture. For example, the art and literature of this period show an increasing aesthetic appreciation for wildness, which had previously been castigated as ugly, as well as new sympathy for peripheral experiences and points of view, including those of animals, as well as of the poor and of the human inhabitants of exotic territories.

In the less elevated sphere of the home, more and more people—especially members of the middle classes, which had especially benefited from the advances of the eighteenth century—indulged in sentimental attachment to pets as it became clear that they represented a nature that was no longer threatening. There are many indices of the increase in companion animals, especially dogs, from about the beginning of the nineteenth century. None of these indices is individually conclusive, but together they seem persuasive in their chronological location of the transition. For instance, the returns from the dog tax, originally imposed as a revenue measure during the Napoleonic wars, grew steadily. Publishers discovered a new market for books about dogs where none had previously existed; in addition, periodicals devoted generally to country sports began to feature dogs more prominently. Finally, well into the Victorian period, the official institutions of dog fancying appeared. The first formal dog show was held in Newcastle in 1859; the Kennel Club was founded in 1873, to be followed by a host of clubs devoted to individual breeds; and the first canine *Stud Book* appeared in 1874. Parallel organizations for cats appeared within a few decades (Ritvo, 1987).

Although it was pervasive, the shift in attitudes that encouraged the expansion of pet ownership among ordinary people was not conscious. That is, when people, either in the nineteenth century or, for that matter, at more recent periods, decided to acquire pets, they did not do so with the conscious intention of reenacting a scenario of human conquest and control of nature. Instead, the voluminous Victorian literature of pet-keeping was saturated with the sentimentality characteristic of the period; at least on the surface it was a literature of love rather than one of domination. But this love presupposed, and even celebrated, a satisfactory resolution of the struggle between nature and civilization. Indeed, not only did the conquest of external nature make it possible for many people to own pets, but the safe, captive, and loyal pet reciprocally symbolized the appropriate relation between humans and nature under the new dispensation. One of the underlying attractions of pet ownership may have been the opportunity it offered people to express those unacknowledged or even subconscious understandings. An examination of a few of the standard concerns of enthusiastic nineteenth-century pet owners reveals that beside the rhetoric of affection and admiration in which they routinely described their relationships with their animals ran another rhetoric, one expressed in action as well as in language, that was explicitly concerned with power and control.

For example, many pet-keepers were producers as well as consumers of animals. The whole enterprise of maintaining and improving breeds embodied a metaphorical assertion of domination; the breeder assumed an almost godlike role in planning new variations. And on a more literal level, any attempt at breeding, at least unless pet owners were content to abandon their aspirations and let their animals be guided entirely by their own inclinations, immediately provoked a contest between human will and natural proclivities. On the most literal level, enforcing a predetermined choice of mate required close physical control of one's animal. Thus the fact that it was impossible to disengage dogs that managed to evade attempted restrictions was particularly troublesome, because it could produce a prolonged and public display of resistance to human authority. "Where they are permitted to run about and appear in such a state before the habitations of the respectable ... it is a most disgusting shameful spectacle," objected one early nineteenth-century chronicler, who was motivated by disciplinary as well as prudish concerns; he continued, "there is, perhaps, no nuisance that stands more in need of compulsive correction" (Taplin, 1803). To prevent such unedifying performances, bitches had to be locked up when they were in heat. Like masters whose moral standards for their families and servants were too lax, owners who neglected this responsibility failed in their duties to the community. (This was not the only way in which the exercise of authority over the natural world became metaphorically confounded with dominion within the human sphere.) The "quantities of bastards [canine bastards were meant here, but the ambiguity was not completely accidental], and the dwindled breed of Pointers and Setters" could be laid to the account of pet owners who exercised insufficient surveillance (Thornhill, 1804).

And physical isolation was not enough; if possible, bitches' sentiments and imaginations had to be controlled as well. An eminent Victorian dog fancier recalled a Dandie Dinmont terrier whose wayward emotions made her useless for breeding; she "became enamoured with a deerhound, and positively would not submit to be served by a dog of her own breed." Even bitches who were more compliant might defeat their owners' purposes, if they were allowed so much as to look at attractive dogs of different breeds. Delabere Blaine, who was sometimes known as "the father of canine pathology," had a pug bitch whose constant companion was a white spaniel. He claimed that all her litters were sired by pugs, and all consisted of undeniably pug puppies—but that in tribute to her infatuation one puppy in each batch was white, a color that was not desirable in pugs (Dalziel, 1879).

If ensuring that pets consorted only with their chosen

mates represented a victory over nature on its own terms—that is, overpowering it by brute force—the kind of manipulation involved in deciding which animals to pair exemplified a triumph of human intelligence. Thus aficionados of many modern dog breeds point with pride to their dogs' ancient origins, but in fact almost all of these origins were fabricated by nineteenth-century dog fanciers in search of distinctions that could be ratified by an elaborate hierarchy of pedigrees and dog show awards. Even the few breeds for which there is significant pre-nineteenth-century evidence, such as the toy spaniel, were transformed so radically by Victorian breeders as to practically obliterate their genetic connection with their alleged forebears. Most so-called traditional breeds simply did not exist as breeds in the modern sense of race or strain at all. For example, the word "bulldog" is old, but until the nineteenth century it referred to dogs that performed a particular function—that is, attacking tethered bulls—rather than to dogs that shared a particular ancestry or set of physical characteristics (Lytton, 1911; Ritvo, 1986). Thus Victorian dog fanciers were working with a relatively clean slate when they set out to develop breed standards. And the standards they developed suggest that what they valued was arbitrariness—the ability to produce animals with surprising or unnatural characteristics. Often the rarest traits, meaning those that were furthest from a strain's inherent inclinations and so offered the strongest evidence of the breeder's influence, provoked the greatest admiration; conversely, typical animals—that is those that displayed traits that would probably have been manifest without human interference—were frequently dismissed as merely mediocre.

Fortuitously, as the most plastic of domesticated animals, dogs were particularly vulnerable to this kind of manipulation; indeed, it was occasionally suggested that their genetic malleability was the gratifying physical analogue of their temperamental eagerness to serve their human masters. Thus, the inclination to celebrate animals that exemplified the human ability to reconstruct had an influence on the character of breed after breed. For example, until bull-baiting was outlawed in 1835, bulldogs were simply dogs of any appearance and ancestry with sufficient courage, strength, and ferocity to hold their own against outraged bulls. It is not surprising that bulldogs of this sort became uncommon once their primary function was abolished; but it may be somewhat surprising that animals of the same name reappeared as popular pets toward the end of the century. However, the terms in which the resurgent bulldogs were described strongly suggested that their connection with their bloodthirsty namesakes was rather tenuous. They were said to be more pampered than other breeds, more "delicate" as puppies, and so indolent that it was necessary to coax them to eat. In disposition, they were alleged to be "peaceable" and "intelligent," also qualities for which their predecessors were seldom celebrated (Anon., 1894; Davies, 1905; Lane, 1900; Pybus-Sellon, 1885). In physical character, the revived bulldogs also reflected the arbitrary manipulations of breeders rather than any adaptation to function. So apparently random were the standards prescribed for the breed that outsiders had great difficulty in making sense of them. Thus, although much printed advice about bulldog breeding was available, a correspondent who said he had "only quite recently entered the Bulldog Fancy" implored the editors of the *Sportsman's Journal and Fancier's Guide* to publish a brief description of "the points, general make and shape ... of the bulldog" (Anon., 1879).

A sample of what this gentleman was up against was the Dudley nose question, which convulsed the Bulldog Club for over a decade. Dudley, or flesh-colored, noses occurred in some strains of fawn-colored bulldogs, usually in conjunction with light eyes and a yellowish face.

In 1884, the Club voted to exclude all dogs with Dudley noses from competition, narrowly defeating a counter-proposal that Dudley noses be considered mandatory in fawn bulldogs (Farman, 1898). This issue was fought on aesthetic grounds, but breeders could be similarly whimsical in their selection of traits that had more serious pragmatic implications. Thus in the 1890s the conformation referred to as being "well out at the shoulder" became standard among well-bred bulldogs. Any dog lacking this feature was doomed to mediocrity in the eyes of show judges, but those lucky enough to display it were likely to end up as seriously crippled as Dockleaf, a renowned champion of this period, who could not walk two miles without collapsing (Anon., 1891a; Lee, 1893).

Almost every dog breed provided occasion for this kind of arbitrary display of human ability to manipulate. The collie, which Queen Victoria's partiality made the most popular late-nineteenth-century dog breed, was also reconstructed to serve the figurative needs of fanciers. Collies were originally valued for the qualities they had developed as hard-working Scottish sheepdogs—intelligence, loyalty, and a warm shaggy coat. But once they were ensconced in the *Stud Book* and in comfortable homes, breeders began to introduce modifications and "improvements." As pedigreed collies became more numerous, breeding fashions became more volatile; breeders redesigned their animals and restocked their kennels in accordance with the latest show results. For example, the 1890s saw a craze for exaggerated heads with long, pointy noses, despite the objections of some conservative critics. In 1891, a *Kennel Gazette* reviewer complained that show judges had given all the prizes to "dogs of this greyhound type whose faces bore an inane, expressionless look." Others alleged that such dogs could hardly display the intelligence characteristic of their breed because there was "no room in their heads for brains" (Anon., 1891b).

Thus breeding offered dog owners the chance to stamp canine raw material with designs of their own choosing; it was a continually repeated symbol of the human ascendancy over nature. (Cat fanciers tried this too, but feline raw material proved much less pliable. This may explain, at least in part, the relatively limited popularity of pet cats during the nineteenth century.) And the theme of control surfaced more explicitly in connection with other pet-related issues. The relatively close co-existence of a large animal population and a large human population inevitably produced conflicts and problems that could only be resolved by regulation. But such regulation often seemed to address issues much more complex and far-reaching than was required merely to eliminate (usually canine) nuisances. Disciplining animals could be confounded with disciplining people, and the regulation in question might easily become the occasion for an unacknowledged redefinition of the boundaries of civilized or responsible society. Unreliable or inadequately disciplined groups of human beings could be grouped with animals rather than with their more respectable conspecifics; thus they frequently replaced nature as the object of manipulation.

Throughout the nineteenth century, for example, even as pets were made increasingly welcome at respectable domestic hearths, the pets of the poor were castigated as symbols of their owners' depravity—an unwarranted indulgence that led them to neglect important social duties. A typical complaint criticized colliers who "have more dogs than they know what to do with" and "starve their children and feed their dogs on legs of mutton." In addition, pet dogs were alleged to intensify the squalor of impoverished accommodations. An eminent veterinarian painted a distressing picture that conflated physical and moral contamination: "currish brutes ... living with their owners in the most miserable and badly ventilated dwellings...and contributing to make these dwellings still more insalubrious by absorbing their share of the oxygen ... and poisoning the atmosphere by their filthiness" (Anon., 1897; Fleming, 1872). Thus reformist critics presented their efforts to deprive the poor of their pets as straightforward humanitarian efforts on behalf of suffering people and animals. But the juxtaposition of such efforts with explicit attempts to regulate the behavior of lower class humans suggests an additional dimension. For example, the regulations governing the Peabody Model Dwellings, part of a paternalistic late Victorian scheme to assist the worthy poor, combined prohibitions against keeping dogs with similar strictures forbidding hanging out laundry, papering the walls, and children playing in the corridors. By defining pets as an inappropriate luxury, which the poor had neither the financial means to support nor the moral means to control, more respectable members of society may have implied that pet-keeping was presumptuous for members of the lower classes (Jones, 1984). The underlying symbolism of domination may have defined pet ownership as the prerogative only of those whose social position justified some analogous exercise of power over their fellow human beings.

These examples and speculations may help explain why large numbers of ordinary people did not begin to keep pets until something under two hundred years ago. With only slight extrapolation, they may also suggest why England, beginning in the last century, and the United States in the current century have been distinguished for the number of pets cherished by their citizens as well as for the generosity with which many of these companion animals are treated. Along the same lines, it is not surprising that the protection of wild animals first emerged as an issue in late Victorian Britain and attracts its most energetic and sustained contemporary support in North America and Europe (Doughty, 1975; Fitter and Scott, 1978). The concept of pet is not inevitably limited by species; pets do not even have to share our domiciles. Pets can also be understood as animals to which we maintain a certain relationship of domination mixed with responsibility and generosity (Tuan, 1984). So defined, pets may prove to be nearly ubiquitous. We have now entered a new transitional period in our relation to the natural world, when we exercise power to an extent undreamed of at the beginning of the nineteenth century; most animals, not just those we have chosen to domesticate, depend upon us for their very existence.

References

Anonymous. 1879. *Sportman's Journal and Fancier's Guide: Stud & Stable & Kennel & Curtilage 129*, Feb. 15.
Anonymous. 1891a. Review of the Past Year. *Kennel Gazette* 12:5–6.
Anonymous. 1891b. Collies. *Kennel Gazette 7*:7–8.
Anonymous. 1894. Dogs up to Date: The Bulldog. *Dogs 1*:43.
Anonymous. 1897. Report of the Departmental Committee to Inquire into and Report upon the Working of the Laws Relating to Dogs. *Parliamentary Papers*, c.8320 and c.8378, *XXXIV*:63–4.
Bewick, T. 1975. *A Memoir of Thomas Bewick, Written by Himself*, ed. I. Bain, 1862. Oxford: Oxford Univ. Press.
Caius, J. Undated. *Of English Dogges: The Diversities, the Names, the Nature, and the Properties*, trans. Abraham Fleming, p. 13. London: A. Bradley.
Chaucer, G. *The Canterbury Tales: General Prologue*, Fragment 1, Group A, lines 146–47.
Dalziel, H. 1879–80. *British Dogs: Their Varieties, History, Characteristics, Breeding, Management and Exhibition*. London: Bazaar office.
Davies, C. J. 1905. *The Kennel Handbook*. London: John Lane.
Doughty, R. W. 1975. *Feather Fashions and Bird Preservation: A Study in Nature Protection*. Berkeley: Univ. of California Press.
Empson, W. 1951. *The Structure of Complex Words*. Ch. 7–8. New York: New Directions.

Farman, E. 1898. The Bulldog Club. *Kennel Gazette* 19:471.

Fitter, R., and P. Scott. 1978. *The Penitent Butchers: 75 Years of Wildlife Conservation.* London: Fauna Preservation Society.

Fleming, G. 1872. *Rabies and Hydrophobia: Their History, Nature, Causes, Symptoms, and Prevention.* London: Chapman and Hall.

French, R. D. 1975. *Antivivisection and Medical Sciences in Victorian Society.* Princeton Univ. Press.

Hamilton, A. In press. *History of the Ancient and Honorable Tuesday Club,* ed. R. Micklus. Chapel Hill: Univ. of North Carolina Press.

Harwood, D. 1928. *Love for Animals and How It Developed in Great Britain.* New York: Columbia Univ. Press.

Hibbert, C., ed. 1984. *Queen Victoria in Her Letters and Journals,* p. 205. London: John Murray.

Hobsbawm, E., and T. Ranger, eds. 1983. *The Invention of Tradition.* Cambridge: Cambridge Univ. Press.

Jesse, E. 1835. *Gleanings in Natural History: Third and Last Series,* p. vi. London: John Murray.

Jones, G. S. 1984. *Outcast London: A Study in the Relationship between Classes in Victorian Society,* p. 186. New York: Pantheon.

Kellert, S. R. 1983. Affective, Cognitive and Evaluative Perceptions of Animals. In *Behavior and the Natural Environment,* eds. I. Altman and J. F. Wohlwill, 241–67. New York: Plenum Publishing Co.

Lambton, L. 1985. *Beastly Buildings: The National Trust Book of Architecture for Animals,* pp. 165, 168. London: Jonathan Cape.

Lane, C. 1900. *All About Dogs: A Book for Doggy People,* pp. 179, 183. London: John Lane.

Lansbury, C. 1985. *The Old Brown Dog: Woman, Workers, and Vivisection in Edwardian England.* Madison: Univ. of Wisconsin Press.

Lee, R. B. 1893–4. *A History and Description of the Modern Dogs of Great Britain and Ireland* 3:243. London: Horace Cox.

Lytton, J. N. 1911. *Toy Dogs and Their Ancestors, Including the History and Management of Toy Spaniels, Pekingese, Japanese and Pomeranians.* London: Duckworth.

Passmore, J. 1975. The Treatment of Animals. *Journal of the History of Ideas* 36:195–218.

Paulson, R. 1979. The English Dog. In *Polite Art in the Age of Hogarth and Fielding*, 54–6. Notre Dame, IN: Univ. of Notre Dame Press.

Pybus-Sellon, J. S. 1885. Bulldogs. *Kennel Gazette* 5:144.

Ritchie, C. I. A. 1981. *The British Dog: Its History from Earliest Times*, 118–19. London: Robert Hale.

Ritvo, H. 1986. Pride and Pedigree: The Evolution of the Victorian Dog Fancy. *Victorian Studies* 29:227–53.

Ritvo, H. 1987. *The Animal Estate: The English and Other Creatures in the Victorian Age*, ch. 3. Cambridge: Harvard Univ. Press.

Serpell, J. 1986. *In the Company of Animals: A Study of Human-Animal Relationships*. Oxford: Basil Blackwell.

Shell, M. 1986. The Family Pet. *Representations* 15:121–53.

Stillman, W. J. 1899. A Plea for Wild Animals. *Contemporary Review* 75:674.

Taplin, W. 1803. *The Sportsman's Cabinet* 1:27–8. London.

Thomas, K. 1983. *Man and the Natural World: A History of the Modern Sensibility*, p. 112 and ch. 3. New York: Pantheon.

Thornhill, R. B. 1804. *The Shooting Directory*, p. 35. London.

Tuan, Y-F. 1984. *Dominance and Affection: The Making of Pets*. New Haven: Yale Univ. Press.

Turner, J. 1980. *Reckoning with the Beast: Animals, Pain and Humanity in the Victorian Mind*. Baltimore: Johns Hopkins Univ. Press.

Pet-Keeping in Non-Western Societies: Some Popular Misconceptions

James A. Serpell

While Ritvo analyzed the origins of pet-keeping in Western society, Serpell discusses the practice in other cultures. He comments that throughout history, the world's wealthy and ruling classes have demonstrated a powerful affinity for pets. In the modern West, the recent growth of pet populations has coincided with rising standards of living. This apparent association between pet-keeping and material affluence has helped to create the impression that pet-keeping is a practice seen in affluent cultures, a leisure activity of little or no social significance.

The assumption that companion animals serve no useful purpose is prevalent in the field of anthropology. Although the practice of capturing, taming, and keeping wild animals for companionship is widespread among hunting and gathering and simple horticultural societies, it has only rarely been studied or even described in any detail, and explanations for its existence are often strangely contrived. Admittedly, a certain confusion surrounds the meaning of the term "pet." Social anthropologists and historians have undoubtedly devoted considerable attention to the use of animals as adornments, emblems of status, religious symbols, or even as educational "toys." The word "pet" has been applied in each case. They have not, however, managed to explain satisfactorily why so many non-affluent cultures nurture and cherish companion animals without any obvious ulterior motives in mind. Indeed, they have tended to evade the issue by turning it on its head. Rather

*than tackling the reasons why such societies should keep compan-
ion animals at all, they have addressed, instead, the question of
why these societies do not kill them and eat them—as if the only
sensible reason for keeping an animal is in order, ultimately, to
devour it.*

*Research in other disciplines within the last fifteen or so years
has begun to shed light on the potential social, emotional, and recreational value of companion animals in human society. Recognition
of the fact that pets are not, after all, entirely useless may help to
promote a more open-minded approach to what is a fascinating
and, alas, fast vanishing aspect of tribal culture.*

*Serpell demonstrates by example how illuminating a cross-cultural
and multidisciplinary approach can be.* EDITOR

Popular beliefs and misconceptions about why people
keep pets take a variety of forms. Probably the most
widespread is the idea that pets are merely ersatz and,
by inference, inferior replacements for human relationships, and that the people who keep pets must therefore
be, in some way, socially or emotionally inadequate.
The perception of pets as "child substitutes" is also
roughly subsumed by this theory. Another even more
disparaging view of pets sees them essentially as an
artful collection of social parasites who inveigle themselves into human affections by manipulating and sub-
verting our so-called parental instincts. By implication,
then, pet owners are the victims of their pets in the
same sense that sick people are often the victims of disease (Serpell, 1983, 1986). Finally, there is the belief that
pet-keeping is basically a pointless and unnecessary
luxury; a mere by-product of Western wealth, which,
while not directly harmful, is nevertheless wasteful in
terms of emotional and material resources. The present
paper explores the historical origins of this latter idea,
and re-examines some of the erroneous assumptions
upon which it is based.

The Historical Links between Pet-keeping and Wealth

In Europe since classical times there has been an apparent class distinction between those who did and those who did not keep animals as pets. It is clear, for example, that the gentry and nobility of ancient Greece kept pets, since the practice was the subject of a certain amount of popular satire at the time. One of the fictional characters invented by the author Theophrastus (372–278 B.C.) kept monkeys and apes, a Maltese dog, and a tame jackdaw for whom he purchased various toys and accessories. According to Plutarch the Athenian aristocrat, Alcibiades, once paid 70 minae for a dog—more than 20 times the value of a human slave—whose long and gorgeous tail he cut off merely to shock people (Halliday, 1922). The early Greek inhabitants of Sybaris in southern Italy, whose name has since become a byword for luxury and opulence, were also besotted with lapdogs, taking them to bed with them and carrying them about wherever they went, even to public baths. Like the Greeks, the Roman upper classes were also extravagantly fond of their companion animals. The poets Ovid, Catullus, and Martial all wrote lyrical verses in praise of people's pets; the Emperor Hadrian buried his favorite dogs beneath monumental tombstones, and the daughter of Drusus adorned her pet turbot—a kind of flatfish—with gold rings. Not to be outdone, the orator, Hortensius, apparently burst into tears when his turbot suddenly expired (Halliday, 1922; Merlen, 1971; Penny, 1976).

From the Middle Ages onward we find much the same sort of thing; the aristocracy and the ecclesiastical elite lavishing attention on their pets while largely ignoring the unenviable plight of the ordinary working population. Thomas à Becket and many other senior clergymen, for instance, frequently kept dogs and monkeys in their chambers, and we are informed by one chronicler of the period that this was the custom among prelates "for occasionally dispelling their anxieties" (Labarge, 1980). Convents and nunneries were often overrun with "birds, rabbits, hounds and such like frivolous creatures" to which, according to William of Wykeham, the nuns "gave more heed than to the offices of the Church." Often these monastic pets belonged to aristocratic ladies who lived for various periods of time within convents (Ritchie, 1981). Throughout medieval Europe, lapdogs and cats which were of little, if any, utilitarian value were kept in most baronial households. Noble ladies carried them about in their arms and fed them with morsels of food from the table; a habit deplored by contemporary writers on etiquette who vainly insisted that it was impolite to fondle animals at mealtimes (Labarge, 1980). By the sixteenth century, lap dogs were all the rage among the upper crust of English society. In his commentary in Holinshed's *Chronicles of England, Scotland and Ireland*, William Harrison describes these dogs, somewhat sarcastically, as:

> *little and prettie, proper and fine, and sought out far and neere to satisfie the nice delicacie of daintie dames, and wonton women's willes; instruments of follie to plaie and dallie withall, in trifling away the treasure of time (Jesse, 1866).*

Mary Stuart, also known as Mary Queen of Scots, may have played an important part in setting contemporary pet-keeping trends by surrounding herself with an entourage of tiny dogs, some of whom she dressed in blue velvet suits to keep them warm in winter. She was so attached to at least one of these animals that she went to the scaffold with it carefully concealed beneath her petticoats (Jesse, 1866; Szasz, 1968). She also founded an entire dynasty of dog-loving monarchs who ruled Britain for over a century. Her son James I, his son Charles I, and Charles's three children, Charles II, James I, and Mary, were all enthusiastic dog owners. Indeed,

Charles II's fondness of dogs, particularly the little spaniels that now bear his name, was almost as notorious as his exploits with the ladies. Dogs overran the palace during his reign, inspiring one courtier to remark, "God save your Majesty, but God damn your dogs" (Ritchie, 1981). The royal pet-keeping tradition has, of course, been maintained ever since. Queen Victoria had many dogs, including a pair of Pekingeses sent to her by the Dowager Empress of China, and the present monarch, Elizabeth II, is world-famous for her ever-present coterie of corgis.

Pet-keeping among the ruling classes was not by any means a purely European phenomenon. For at least a thousand years, the Emperors of China, for example, kept dogs and, less often, cats in their royal apartments, and ennobled them with the ranks of senior court officials. Under the Manchurian Ch'ing Dynasty the ancestors of the modern Pekingese enjoyed a privileged status unrivaled by any other variety of pet before or since. They were given the titles of princes and princesses, and huge personal stipends were set apart for their benefit. As puppies they were suckled at the breasts of human wet nurses, and as adults they were attended by a retinue of hand-picked servants. A special elite corps of royal eunuchs was also created to supervise their overall care and husbandry (Dixie, 1931). Japan also had its fair measure of dog-loving rulers. During the seventeenth century, one Shogun became so enthusiastic that he provided food and shelter for about 100,000 dogs. The cost of caring for these pets overburdened the national Exchequer, produced inflation, and resulted in an unpopular new tax on farmers (Watts, 1985). Even Africa was not exempt from this form of extravagance. When John Hanning Speke visited Uganda in 1862, he found the palace of King M'tesa infested with pets of every description. The King himself was particularly fond of a small white dog, which followed him around attached to a piece of string (Speke, 1863).

Figure 1. King M'tesa of Uganda walking his dog (From: the original drawing by J. H. Speke. 1863. Journal of the Discovery of the Nile, *292. London: Blackwood.)*

During the course of the last century, pet-keeping has gradually achieved full emancipation in the Western world, and ownership of companion animals is now fairly evenly distributed across all social classes (Messent and Horsfield, 1985). But, again, this proliferation of pets in modern industrial societies has been accompanied by a steady increase in human living standards, and many would argue that this is sufficient evidence on its own that pet-keeping is a mere by-product of Western affluence; a self-indulgent waste of emotional and material resources that would be better spent in service of underprivileged human beings (see Szasz, 1968; Baxter, 1984). This view of pets has been around for a considerable period of time.

The Roman writer Plutarch, for instance, was among the first to voice his disapproval of pet-keeping in precisely these terms:

> *Caesar once, seeing some wealthy strangers at Rome, carrying up and down with them in their arms and bosoms young puppy dogs and monkeys, embracing and making much of them, had occasion not unnaturally to ask whether the women in their country were not used to bear children; by that prince-like reprimand gravely reflecting upon persons who spend and lavish upon brute beasts that affection and kindness which nature has implanted in us to be bestowed on those of our own kind (Halliday, 1922).*

Similarly, when William of Wykeham criticized the nuns of Romsey Abbey for keeping pets in 1387, he noted that these animals were devouring alms which should have been given to the poor (Ritchie, 1981). William Harrison, writing in the sixteenth century, was more blunt. He described the nobility as wanton, idle, and corrupt because of their pet-keeping activities, and he then went on to deliver a scathing attack on "people who delight more in their dogs that are deprived of all possibilitie of reason, than they do in children that are capable of wisdome and judgement. Yea, they oft feed them of the best, where the pore man's child at their dores can hardlie come by the worst" (Jesse, 1866). Moral diatribes of this kind against pets did not fall entirely on deaf ears. According to one account, a pious Elizabethan lady called Katherine Stubbes deeply repented all the affection she had shown her pet dog. On her deathbed she is reported to have said to her husband:

> ... *you and I have offended God grievously in receiving many a time this bitch into our bed; we would have been loath to have received a Christian soul ... into our bed, and to have nourished him in our bosoms, and to have fed him at our table, as we have done this filthy cur many times. The Lord give us grace to repent it (Thomas, 1983).*

In other words, because of its apparent association with wealth and social inequality, pet-keeping has unwittingly become one of the more potent symbols of man's inhumanity to man; conjuring up visions of villainous and despotic rulers doting over plump little lap dogs while their unfortunate subjects perished from neglect, starvation, and disease.

The issue is clearly an important and emotional one, but there is a grave danger of allowing such powerful images to distort our perceptions of the whole phenomenon. The assumption that pet- keeping is a trivial and wasteful spin-off of material wealth rests on the notion that poor or non-affluent people do not keep pets. Even in Europe this was not always the case. During the sixteenth and seventeenth centuries pet-keeping was probably relatively commonplace among the poorer classes, although whenever it was detected it aroused grave suspicions. At the time of the English witch trials (1570–1700), for example, the possession of an animal pet or "familiar" was frequently used as evidence for accusations of necromancy, and most of the victims of this persecution were elderly and financially impoverished

Figure 2. English witches and their familiars. (From: "The Wonderful Dis-coverie of the Witches of Margaret and Phillip Flower," 1619.)

(Serpell, 1986). This antipathy for pets was certainly not motivated by any economic considerations. It arose because, at the time, affectionate relationships between people and animals were regarded as immoral. Indeed, one moralist of the period explicitly condemned "over familiar usage of any brute creature," presumably out of the curious but popular conviction that such intimate contacts with animals could somehow brutalize or dehumanize people (Thomas, 1983). Elsewhere in the world, the links between pet-keeping and obvious symptoms of material affluence were even more tenuous than they were in Europe.

Pet-Keeping in Tribal Societies

When European explorers first set out to investigate the uncharted regions of the world between the sixteenth and nineteenth centuries, they were generally astonished to find the homes and villages of the native inhabitants infested with pets of every description. Early accounts of the Indians of North America, for example, describe how these peoples kept tame raccoons, moose, bison, wolves, bears, and innumerable other species as pets, and how they loved and fondled their dogs with every sign of affection (Hernandez, 1651; Galton, 1883; Linton, 1936; Elmendorf and Kroeber, 1960; Mooney, 1975). The relationship between the Indians and their companion animals does not appear to have been fundamentally different from that which we associate with the modern West. Writing in the eighteenth century, for instance, Sir John Richardson noted that "the red races are fond of pets and treat them kindly; and in purchasing them there is always the unwillingness of the women and children to overcome, rather than any dispute about price" (Galton, 1883). He also observed that the women gave their bear cubs milk from their own breasts—not a practice one sees very often in Western societies!

In South America, animal-taming and pet-keeping were even more popular. Two early Spanish explorers reported that although the Indian women kept tame birds and animals in their huts:

> ... they never eat them: and even conceive such a fondness for them that they will not sell them, much less kill them with their own hands. So that if a stranger who is obliged to pass the night in one of their cottages, offers ever so much for a fowl, they refuse to part with it, and he finds himself under the necessity of killing the fowl himself. At this his landlady shrieks, dissolves into tears, and wrings her hands, as if it had been an only son (Juan and Ulloa, 1760).

The list of animals tamed and kept by these Indians covered virtually all of the common birds and mammals available to them. The nineteenth-century English naturalist, Bates, mentions "twenty-two species of quadrupeds" that he found living tame among the villages of the Amazon basin (Galton, 1883), and the anthropologist Roth (1934) described how the women would "often suckle young mammals just as they would their own children; e.g. dog, monkey, opposum-rat, labba, acouri, deer, and few, indeed, are the vertebrate animals which the Indians have not succeeded in taming." Ironically, the intrusion of Western society and values into South America has brought about a decline in pet-keeping along with the native cultures practicing it. The more remote tribes, however, still retain the habit. The Caraja people of Brazil, whose lands are now threatened by a massive development project, were, according to a visitor in the 1930s, devoted to their pets:

> The villages swarmed with livestock. At nightfall parrots warred with scrawny poultry for roosts along the roof-pole. Pigs, and dismal dogs, and fantastically prolific cats, and tame wild ducks wandered in and out of the huts through holes in the wall. In almost all of the northerly villages cormorants paddled among the litter round the cooking fires; sometimes their sombre plumage had been decorated by the children with tufts of red arara's feathers fastened to their wings (Fleming, 1984).

The Warao, who live around the mouth of the Orinoco River, keep wild birds, monkeys, sloths, rodents, ducks, dogs, and chickens as pets (Wilbert, 1972) and, according to the anthropologist Basso (1973), the Kalapalo Indians of central Brazil maintain a particular affection for pet birds. She describes the relationship between the Kalapalo and their birds as similar to that between human parents and their children. The birds are fed, reared, and protected within the confines of the house, and are often kept in seclusion, like human adolescents "to make them more beautiful." Pet-keeping also remains one of the principal leisure activities of the Barasana Indians of eastern Colombia. Rodents, dogs, parrots, and a huge variety of other large and small birds are the most common pets, although tapir, peccary, ocelot, margay, domestic cats, and even jaguars are also kept in small numbers. The women suckle puppies

and hand-feed other young mammals; they also masti-
cate plant foods such as manioc and banana to feed to
their tame parrots and macaws. One individual was
also observed to spend several hours each day catching
small fish to feed a tame kingfisher. According to the
Cambridge anthropologist, Stephen Hugh-Jones, who
has studied these people for many years, Barasana pet-
keeping is not motivated by any practical or economic
considerations. These people simply enjoy looking after
and caring for their pets. The animals are a continual
source of discussion and entertainment, and are treated
as an integral part of the community (Hugh-Jones, pers.
comm.; Serpell, 1986).

It is important to emphasize that affection for pets
within such societies is largely independent of economic
considerations. Although many of the species kept as
companion animals were also hunted and killed for
food, these same species were exempt from slaughter
once they had been adopted as pets. Referring to the Indians of Guiana, Roth (1934) is quite firm in stating that
"the native will never eat the bird or animal he has
himself tamed any more than the ordinary European
will think of making a meal of his pet canary or tame
rabbit." Such inhibitions were equally strong in societies
where the animal involved was also raised commer-
cially as an item of food. In Hawaii, for example, dogs
were commonly raised for the pot, but pet dogs were
rarely slaughtered or consumed, and never without
loud protests from the owner (Luomala, 1960). Even
when well-intentioned Europeans pointed out the
potential economic uses of pet animals, few of these
cultures took their ideas seriously. The Caraja, for instance, refused to sell some of their pet parrots regardless of how much visitors were prepared to pay for
them. And they treated the whole concept as a joke
when it was suggested that they train their pet cor-
morants to catch fish by fastening rings around their
necks: "In conception, rather than in execution, this

project amused them very much; it is clear that they thought of the birds always as guests, never as servants" (Fleming, 1984).

Yet despite the apparent absence of economic motives, many early explorers and later anthropologists seemed determined to believe that utilitarian considerations were somehow involved. The Swedish explorer Lumholtz (1884), for example, observed that the Australian Aborigines were absurdly fond of their pet dingoes, rearing them:

> ... *with greater care than they bestow on their own children. The dingo is an important member of the family; it sleeps in the huts and gets plenty to eat, not only of meat but also of fruit. Its master never strikes, but merely threatens it. He caresses it like a child, eats the fleas off it, and then kisses it on the snout.*

The only rational explanation he could think of to account for this bizarre (from his perspective) behavior was the fact that the dingo "is very useful to the natives, for it has a keen scent and traces every kind of game." More than eighty years later, anthropologists were attempting to make the same connections. Harrison (1965) states that the Dyaks of North Borneo "literally *love* their dogs" in return for this animal's aid in hunting, and Cipriani (1966) likewise accounts for the Andaman Islander's "inordinate love of dogs" by the fact that dogs meant "invariable and abundant success in the hunt." But clearly, as the plight of modern factory-farmed livestock testifies, mere economic utility provides no guarantee of affection. The B'Mbuti Pygmies of Zaire, for instance, almost invariably hunt with dogs. Yet they have a reputation for treating their canine companions with pointless brutality (Singer, 1968). Conversely, the Comanche of North America were besotted with their dogs, although these animals had no economic value whatsoever (Linton, 1936).

Another popular utilitarian explanation sees pets pri-

Figure 3. Punan Dyak with his dog— affection for these animals is wide-spread in tribal societies (From: Harrison, 1965. Reproduced by Permission of the Council of the Malaysian Branch of The Royal Asiatic Society. RAS Journal 38: 2.)

marily as educational "toys." According to this theory, children who have the opportunity to observe, play, and interact with such animals gain experience that will enable them to become more successful hunters in later life (Laughlin, 1968). This idea appears to stem largely from confusion over the various meanings of the term "pet." It is undoubtedly true that in many hunting societies children tend to be given small wild animals as temporary playthings. Like Christmas gifts in our own culture, these unfortunate, animated toys are usually short-lived, and often end up the objects of target practice or mutilation. It is entirely possible that these childhood games provide practice and instruction for future hunting activities, but it would be a great mistake to confuse this with the kind of animal/human relationships characteristic of the Warao, the Kalapalo, the Barasana or, indeed, most of the cultures already described. The pets in these latter societies are breast-fed, nurtured, protected, and cared for throughout their lives. In no sense can they be regarded as expendable objects of entertainment. The trouble is that the word "pet" covers a multitude of sins, and it is important, whatever the society, to distinguish between companion animals and animals used as objects of play, status, or, indeed, any other purpose.

The subject of pet-keeping in tribal societies has also contributed to an ongoing debate between "structuralist" anthropologists and "cultural materialists" about the origins of dietary and sexual taboos. Structuralists have argued, for instance, that people avoid killing and eating pets because the animals have been personified and included in the social world of people (Levi-Strauss, 1966; Leach, 1964; Sahlins, 1976). A moot point, no doubt, but it entirely fails to explain why they keep the animals or personify them in the first place. Others have pointed out the symbolic resemblance between the act of eating a pet and the act of sexual intercourse between close relatives. According to this view, we don't eat our pets because it would be metaphorically equivalent to committing incest (Tambiah, 1969). Cultural materialists, taking a more down-to-earth perspective, have suggested that the real reason we don't consume companion animals such as dogs and cats is simply because of the practical and economic difficulties associated with farming these carnivorous species for food (Harris, 1978). Neither side in this debate attempts to explain why subsistence hunters and horticulturalists invest so much of their time and resources in economically valueless pet animals; they are solely concerned with people's reluctance to kill pets and eat them. As if the only sensible or understandable reason for keeping and caring for an animal is in order, ultimately, to devour it.

Discussion and Conclusions

There appear to be two main reasons why anthropologists have been reluctant to explore the phenomenon of pet-keeping or to speculate about its functions. Until comparatively recently, attitudes toward so-called primitive societies have been influenced strongly by old-fashioned, ethnocentric views of human cultural development. According to this tradition, societies evolved progressively upward toward increasingly advanced and sophisticated levels of material civilization. Because they were seen as occupying the lowest rungs of this developmental ladder, the lives of hunters and simple horticulturalists were assumed to be correspondingly arduous and uncomfortable. Viewed in this light, hunting economies could not afford to engage in nonproductive activities such as pet-keeping, so the practice was best ignored, explained away as aberrant, or squeezed into some form of contrived utilitarian hypothesis. Fortunately, however, within the last twenty years, ideas about hunting and gathering have changed dramatically. Research on contemporary huntergatherers (see for instance Lee, 1969), and the work of paleoanthropologists and pathologists (such as Cohen and Armelagos, 1984) suggests that subsistence hunters, both now and in the past, often enjoy more leisure time, and are generally healthier and better nourished than many agricultural populations. In other words, hunters and horticulturists appear to be relatively affluent (although not perhaps in the sense that we use the term in the West), and there does not seem to be any economic reason why they should not also keep pets.

Attitudes to pets have also changed. Whereas petkeeping was once assumed to be a pointless luxury or a curious perversion, it can now be understood as the outcome of normal human social behavior and needs. During the last fifteen or so years, the work of Boris Levinson, Sam and Elizabeth Corson, Aaron Katcher and Alan Beck, Leo Bustad, Mike McCulloch, Peter Messent, and many others has amply demonstrated that the majority of pet owners are normal, rational people who make use of animals in order to augment their existing social relationships, and so enhance their own psychological and physical welfare. And in all probability, this is as true for South American huntergatherers as it is for people in the industrial West. Thought of in these terms, keeping a dog, a cat, a parrot, or even a tapir for companionship is no more outlandish or profligate than wearing an overcoat to keep out the cold. This does not, of course, mean that petkeeping is universally beneficial since, like any leisure activity, the net benefits need to be weighed against the costs. It does, however, imply that, where adequate time and resources are available, pet-keeping will arise as a natural and beneficial product of human social propensities.

One of the more attractive aspects of this new concept of human/animal relationships is that it allows us to approach and re-examine many old problems from a novel perspective. Within the field of anthropology, pet-keeping remains virtual *terra incognita* as an area of research. Yet it is one that in the future may provide important insights into, among other things, the origins of animal domestication, the emotional and affiliative needs of non-Western peoples, and the relationship that exists between modes of economic subsistence and overall attitudes towards animals and the natural world (see Serpell, 1986).

It is undeniably true that humans, like all animals, are ultimately constrained by material or, more correctly, ecological demands. But any attempt to understand the evolution of human behavior purely in terms of these *essentials* will inevitably ignore a wealth of social and cultural factors that people may be able to live without, but that nevertheless make a substantial contribution to the quality of their lives. The keeping of animals as companions is clearly not essential to human survival. We can live without it, just as we can live without singing, dancing, music, art, laughter, and friendship. Yet the fact that so many people in so many different cultures are motivated to engage in these inessential activities strongly suggests that the rewards are far from negligible.

References

Basso, E. B. 1973. *The Kalapalo Indians of Central Brazil*, p. 21. New York: Holt, Rinehart and Winston.

Baxter, D. N. 1984. The Deleterious Effects of Dogs on Human Health: 3. Miscellaneous Problems and a Control Programme. *Community Medicine* 6:200.

Cipriani, L. 1966. *The Andaman Islanders*, pp. 80–1. London: Weidenfeld & Nicholson.

Cohen, M. N., and G. J. Armelagos. Paleopathology at the Origins of Agriculture: Editor's Summation. In *Paleopathology at the Origins of Agriculture*, eds. M. N. Cohen and G. J. Armelagos, 585–601. New York: Academic Press.

Dixie, A. C. 1931. *The Lion Dog of Peking*. London: Peter Davies.

Elmendorf, W. W., and K. L. Kroeber. 1960. The Structure of Twana Culture with Comparative Notes on the Structure of Yurok Culture. *Washington University Research Studies,* Monograph 2, *28*:114.

Fleming, P. 1984. *Brazilian Adventure,* pp. 145–46. London: Penguin.

Galton, F. 1922. *Inquiry into Human Faculty and Its Development,* pp. 246–53. London: Macmillan.

Halliday, W. R. 1922. Animal Pets in Ancient Greece. *Discovery* 2:151–54.

Harris, M. 1978. *Cannibals and Kings,* pp. 120–21. London: Collins.

Harrison, T. 1965. Three "Secret" Communication Systems among Borneo Nomads (and Their Dogs). *J. Malay. Branch Roy. Asiatic Soc. 38*:67–86.

Hernandez, F. 1651. Historiae Animalium et Mineralium Novae Hispaniae. In *Rerum Medicarum Novae Hispaniae,* eds. N. A. Recchi and J. T. Lynceus (Rome, 4to), 295–96.

Jesse, G. R. 1866. *Researches into the History of the British Dog.* London: Robert Hardwicke.

Juan, G., and A. de Ulloa. 1760. *Voyage to South America* 1:426. London.

Labarge, M. W. 1980. *A Baronial Household of the Thirteenth Century,* p. 184. Brighton: Harvester Press.

Laughlin, W. S. 1968. Hunting: An Integrating Biobehaviour System and Its Evolutionary Importance. In *Man the Hunter,* eds. R. B. Lee and I. DeVore, 309. Chicago: Aldine Press.

Leach, E. 1964. Anthropological Aspects of Language: Animal Categories and Verbal Abuse. In *New Directions in the Study of Language,* ed. E. H. Lenneberg, 23–63. Cambridge: MIT Press.

Lee, R. B. 1969. !Kung Bushmen Subsistence: An Input-Output Analysis. In *Environment and Cultural Behaviour,* ed. A. Vayda, 47–9. Garden City, NY: Natural History Press.

Levi-Strauss, C. 1966. *The Savage Mind,* pp. 203–16. Chicago: Chicago Univ. Press.

Linton, R. 1936. *The Study of Man: An Introduction,* pp. 428–29. New York: Appleton-Century-Crofts.

Lumholtz, C. 1884. *Among Cannibals,* pp. 178–79. London: John Murray.

Merlen, R. H. A. 1971. *De Canibus: Dog and Hound in Antiquity*, pp. 63–4. London: J. A. Allen.

Messent, P. R., and S. Horsfield. 1985. Pet Population and the Pet-Owner Bond. In *The Human-Pet Relationship*, 9–17. Vienna: IEMT.

Mooney, M. M. 1975. *George Catlin: Letters and Notes on the North American Indians*. New York: Clarkson N. Potter.

Penny, N. B. 1976. Dead Dogs and Englishmen. *The Connoisseur* Aug., p. 298.

Ritchie, C. I. A. 1981. *The British Dog*, pp. 64–119. London: Robert Hale.

Roth, W. E. 1934. An Introductory Study of the Arts, Crafts and Customs of the Guiana Indians. *38th Annual Report of the Bureau of American Ethnology*, pp. 551–56.

Sahlins, M. 1976. *Culture and Practical Reason*, pp. 174–75. Chicago: Chicago Univ. Press.

Serpell, J. A. 1983. What Have We Got Against Pets? *New Scientist*, 13 Oct.

Serpell, J. A. 1986. *In the Company of Animals*. Oxford: Basil Blackwell.

Singer, M. 1968. Pygmies and Their Dogs: A Note on Culturally Constituted Defence Mechanisms. *Ethos* 6:270–79.

Speke, J. H. 1863. *Journal of the Discovery of the Source of the Nile*, pp. 288–92. London: W. Blackwood.

Szasz, K. 1968. *Petishism: Pet Cults of the Western World*. London: Hutchinson.

Tambiah, S. J. 1969. Animals Are Good to Think and Good to Prohibit. *Ethnology* 8:452–53.

Thomas, K. 1983. *Man and the Natural World: Changing Attitudes in England 1500–1800*, p. 40. London: Allen Lane.

Watts, D. 1985. Touching Tribute to Canine Loyalty. *The Times* (London), 28 Feb.

Wilbert, J. 1972. *Survivors of Eldorado: Four Indian Cultures of South America*, pp. 96–7. New York: Praeger.

Health and Caring for Living Things

Aaron Honori Katcher
and Alan M. Beck

Aaron Katcher and Alan Beck have contributed many valuable insights into the nature of the human-animal interaction since they became involved in this field of research approximately ten years ago. In this chapter, they continue to expand our understanding of the human-animal bond, arguing convincingly that the nurturing of living things, whether vegetables, dogs, or babies, might represent an important human need. The human infant is a helpless creature that needs a great deal of care and nurture; it is speculated that nurturing behavior may have conferred evolutionary advantages.

For much of human history, through the hunter-gatherer stage and the shorter agricultural phase, humans had ample opportunity to nurture living beings and fulfill whatever needs they might have in this regard. Within two hundred years, the Industrial Revolution changed the relationship of many people with nature and animals. There was an enormous shift of population from rural, agricultural settings to noisome and crowded cities. Change has been particularly rampant in the twentieth century: for example, in 1910, there were still as many farm workers as factory workers in the United States; today, only about 3% of the work force is engaged in farming. In the modern industrial world, very little time is spent in contact with plants and animals.

We have no idea what the cognitive, emotional, and physiological consequences, if any, of this very rapid change might be. Katcher

and Beck propose that it is probably important that we try to find
out, implying that the current popularity of pets might be a re-
sponse to lack of nurturing opportunities in a modern industrial
society. EDITOR

This paper has a single, simple, central theme: we are
reawakening to the singular importance of contact with
natural surroundings and companion animals because,
unlike any other generation of human beings since the
inception of agriculture, human existence in industrialized societies is deprived of opportunities for nurturing
and for affectionate interchange with others. This deprivation is a result, in part, of the explosive urbanization
of the population complemented by the mechanization
of agriculture and, most recently, the fall in birth rate.

This paper will argue that care of farm animals, pets,
and gardens permits the elaboration of nurturing beyond the raising of human children, and that the extension of the activities of nurturing in both depth and
time have had favorable consequences.

This insight did not arise from the close study of recorded data. It was formed during a hike in the Pyrenees. These are moderately low mountains, six to nine
thousand feet high, but relatively new, geologically
speaking, and deeply carved by glaciation. In the space
of five or six hours one can walk through secondary
forest to alpine meadows to bare scree, where slender
cataracts fall from snow melting in the July heat. Those
walks were journeys through different ways of appre-
hending nature, a nature manifested in: the shifting patterns of light and shade created by one's own motion
through forest; the hypnotic noise and movement of
swiftly flowing water; the microscopic view of the
ephemeral world of alpine flowers, bright butterflies,
and scurrying beetles; the sight of spring displayed, in
space rather than time, near the melting snow fields,
where one can walk in a few steps from full flower to

barely green thrusting buds, to still-frozen earth; the odors of mint, thyme, mold; and, finally, a sweeping vista from the moraine's rock rim, where a sense of the sublime and of the slowness of time is created by the distant unobstructed vision of other peaks.

Those walks generated a feeling of being intact, complete, as if the solid distinct otherness of that natural world had acted as a mirror reflecting myself back to myself. That sense of being intact, distinct, and comfortable in myself crystallized precisely at the moment when the sense of being a separate self was lost in contemplation.

The realization that a solitary retreat into a more intimately absorbing contact with the natural world could create a sense of completeness suggested a new architecture for thinking about the impact of the natural world on human life. In previous writing (Katcher, 1981) it was suggested that pets did not substitute for human relationships, but complemented them. This conclusion followed from the recognition that most pet owners live in intact families containing both children and pets. These pets gave their owners access to a sensual dialogue combining touch, talk, and mutual attention with a superabundance usually not available from other human beings. Contact with pets, parks, and wilderness was a welcome addition to human experience, adding to its complexity and pleasure. The paper should have suggested that contact with the natural world might be a *necessary* part of human development. Perhaps life in a purely human or cultural environment might be a deprived life, a life where survival is always possible, but a life suffering emotional loss, and a failure to realize potential and, perhaps, a loss of vitality and health. The failure in the paper to recognize the essential role of contact with the natural world arose from a failure to consider an appropriate span of human history. An examination of that context revealed the theme of this paper: in no era have human beings been as deprived of nurturing contact with either children or animals as they are now.

Conventional thinking about animal companionship is very much like the conventional psychological wisdom about gentle touch surfacing immediately after the Second World War. At that time it was believed that infants were bonded to their mother by the act of feeding and were eating machines connected to the outside world only through their mouths. Holding and touching were recognized as comforting, but not necessary and perhaps distracting and addicting. Although the healing virtues of touch for adults had been acknowledged for a very long time, its place in normal adult life was not recognized at all—the psychological sciences, in resonance with the general culture, being still too engaged with the exploration of sexuality. It was not until the work of Spitz (1945), Bowlby (1952), and Harlow (1962) that we were able to recognize touch as a vital necessity for infant development. Without touch infants do not develop normally and, if the deprivation is severe enough, they can fail to thrive and die. Recovered infants or infants who experienced less severe deprivation may have persistent impairments in their ability to form close relationships with other human beings. After infancy human beings can exist without affectionate touch, but such a life is always deprived and that deprivation has a cost in mental and physical health.

To understand how contact with the natural world may be a vital part of normal human experience, it's necessary to see how much of the activity of keeping a pet resembles the nurturing given human infants. Once pet-keeping is seen as an extension of human nurturing, its value becomes more obvious. We can best indicate the resemblance between child and pet care by briefly describing the development of our own research. When we first observed that the presence of a pet in the family could have a positive effect on the health of patients with severe coronary artery disease (Katcher and Friedmann, 1980; Friedmann et al., 1980), we immediately thought of the role of physical display of affection in producing that increase in health. Our initial experiments (Katcher, 1981) demonstrated that the touch-talk dialogue with animals was associated with lower blood pressures than dialogue with people. Those early observations have been essentially confirmed and extended by Baun (1985), Grossberg (1986), Jenkins (1986), and Friedmann et al. (1986).

When people did touch and talk to their animals, they used a distinctive style of facial expression and speech that had the form of a physical, if not a verbal, dialogue. Observing a dialogue of words, touch, eye contact, reciprocal grooming, mouthing, and scenting between people and dogs or cats does not arouse our wonder, perhaps because we believe in a "mammalian bond," a common heritage of social existence. It is more thought-provoking to see that kind of reciprocal grooming and affectionate interchange between people and birds, two species that have been pursuing divergent paths of evolution for over 250 million years. Like dialogue between people and dogs or people and cats, this interaction with birds had the following characteristics (Katcher, 1985; Katcher and Beck, 1986):

1. Where possible the head of the person is placed close to the head of the animal
2. The volume of the voice is reduced, sometimes to a whisper
3. The pitch of the voice is raised
4. The rate of speech and length of utterances are decreased
5. There is considerable verbal play with words, combinations of words and sounds, and stress and length of syllables
6. Utterances are terminated with a rising inflection to emphasize or create a question, permitting the crea-

tion of a pseudo-dialogue with the animal. In this dialogue, the person may supply a verbal response for the animal, or some response of the animal's may be used as a reply. Appropriate pauses are inserted in the dialogue to permit such replies.

All of these characteristics of speech have been noted as characteristic of the dialogue between parents and young infants. To this extent, the dialogue between person and pet resembles the play between parent and human infant.

There are, however, certain characteristics of the dialogue between person and pet that make that kind of dialogue distinctly different from "motherese," and indicate that child-oriented dialogue has been modified for use with animals. The most striking differences are found in the facial expressions used, and in the level of excitation conveyed in the dialogue. Parents tend to use exaggerated facial gestures in interacting with children as if they are miming a caricature of emotional expression to aid the child in the recognition of the appropriate facial gestures. In interaction with animals the face is remarkably composed. The brow is usually smooth, the nasal labial fold is flat, and the eyes are partially closed. This appearance of relaxation is accentuated by the character of the smile, which is less pronounced than the smiles used when talking to the experimenter. It resembles the "Madonna" smile with which parents gaze upon sleeping infants.

A second special characteristic of intimate interaction with companion animals is also related to the level of arousal communicated. With young infants, the mother will use arousing styles of speech and facial gestures to capture the child's attention. With companion animals that are free to move actively, increasing the level of arousal will either cause the animal to escape (or less frequently to attack), or will increase the amplitude of the interaction until it no longer has the characteristics of an intimate dialogue but constitutes play instead. Our subjects, who were told to talk to and touch their animals as they usually do, kept the arousal level reflected in their behavior low, to preserve the intimacy of the dialogue, and to keep the animal receptive to gentle touch. The change in facial expression and demeanor had its effects on observers. People were more attractive when engaged with their pets. Their smiles and their aura of peaceful relaxation were contagious. Perhaps these changes are behind the observation (Messent, 1983) that people are approached more often when they are with their pets in public spaces, and are also present in Lockwood's (1983) conclusions that images of people are perceived more favorably when they are paired with images of companion animals. Our preference for images of people paired with animals is translated into practice. People holding children and people holding pets are two of the most popular subjects for home photographers (Ruby, 1983), and handicapped persons assisted by dogs are approached for social contact much more frequently than when they are out alone (Eddy, Hart, and Boltz, 1986).

Because the dialogue between people and their pets resembles the dialogue between parents and young infants, and because people commonly speak of their pets as members of the family and as children (Cain, 1983; Katcher, 1981), nurturing must be a dominant theme in our relationships with pets. Examination of the general characteristics of nurturing (Table 1) reveals broad similarities in the activities and feelings directed toward children, pets, farm animals, and even gardens and house plants. The table is descriptive and does not make implications about value. A variety of feelings is evoked by the practice of abortion, infanticide, disposal of unwanted pups or kittens by breeders, or slaughter of animals for food, yet all of these activities occur in nurturing relationships. In similar fashion, "domina-

Table 1. Characteristics of Nurturing

Companionship	Domination—Protection	Desirable Consequences
Eating together	Shaping of form	Immersion in cyclical time
Sleeping together	Management of sexuality	Feelings of safety, completeness, control and intimacy
Petting	Constraint of movement	Knowledge
Play	Shift from nature to culture	Relaxation through outward direction of attention
Face-to-face interaction	Working	
Reciprocal grooming	Killing	
Dialogue	Management of excrement	

tion" is a term used pejoratively to describe the attitude toward nature inherent in a consumption-oriented society. Yet it is not possible to raise a well-behaved pup, bird, or child unless they are effectively subordinated. Domination may, of course, have more than instrumental uses; Tuan (1984), by comparing the symbolic structure of pets, gardens, and fountains, has cogently argued that these adornments to human life express a delight in the domination of nature, and in the transformation of the natural into the cultural.

In this era of intensive farming it is difficult to remember that sensuous affectionate relationships exist between farmers and their animals. Fortunately, Hardy has given us an excellent description:

> In general the cows were milked as they presented themselves, without fancy or choice. But certain cows will show a fondness for a particular pair of hands, sometimes carrying this predilection so far as to refuse to stand at all except to their favorite, the pail of a stranger being unceremoniously kicked over. ...
>
> Tess, like her compeers, soon discovered which of the cows had a preference for her style of manipulation. ... Out of the whole ninety-five there were eight in particular—Dumpling, Fancy, Lofty, Mist, Old Pretty, Young Pretty, Tidy, and Loud—who, though the teats of one or two were as hard as carrots, gave down to her with a readiness that made her work on them a mere touch of the fingers. ...
>
> All of the men and some of the women, when milking, dug their foreheads into the cows and gazed into the pail. But a few, mainly the younger ones, rested their heads sideways. This was Tess Durbeyfield's habit, her temple pressing the milcher's flank her eyes fixed on the far end of the meadow with the quiet of one lost in meditation.

THOMAS HARDY, *Tess of the D'Urbervilles*
Penguin Edition, 1984, p. 177

It must also be emphasized that most of the activities and consequences of pet- or livestock-keeping are part of cultivating and gardening. The two activities do differ, however; with pets one is always hauling away excrement while with gardens one is always hauling it in. Nurturing is a biological activity, with touch, odor, rhythmic and cyclical activities playing a large role. One would expect psychological and biological effects from such an activity. Unfortunately, there is little information about the favorable or unfavorable biological consequences of nurturing activity. To understand what domestication of plants and animals may have done for us, it is necessary to look at that activity in an evolutionary context.

It is a reasonable hypothesis that the prolonged care of infants in primate groups was facilitated and maintained by deeply rooted physiological, psychological, and social rewards. If child-rearing makes parents healthier and more socially attractive, the survival of their infants would be facilitated. Survival of progeny would also be facilitated if animals no longer raising offspring were less attractive, and more vulnerable to disease and death. If you wish to ensure the passage of your genes into the next generations, then stay healthy as long as you can, aid your children in the competition for resources, and then pass out of the picture rapidly when you become only another competitor.

When humans give up caring for others and lose their appetite for work and pleasure, they do become more vulnerable to disease. There is a large literature suggesting that social isolation, loss of a spouse, and depression can result in decreased health and significantly increased vulnerability to accident, chronic disease, and death (Lynch, 1977; Ory, 1983). Depression is a complex psycho-biological state in which the competence of the immune system is decreased and mortality from a broad range of pathological events is increased. One of the most important triggers for depression is loss of the opportunity to care for, nurture, and love others. Depression can be triggered by withdrawal from the company of others and, in turn, depression causes increasing social withdrawal. Caring and depression are both self-facilitating states, one causing us to move toward other people and health, and the other leading us to increasing social isolation and vulnerability to disease. The history of human evolution is a history of increasing time spent in nurturing infant animals. The continuing enlargement of the primate brain demanded one of two evolutionary adaptations: the production of progressively altricial young, so that growth in brain size was completed after birth, or massive enlargement of the pelvic outlet. Apparently human development was associated with progressive neoteny, the first of these strategies. As described by Gould (1981):

Flexibility is the hallmark of human evolution. If humans evolved as I believe, by neoteny, then we are, in a more than metaphorical sense, permanent children. (In neoteny, rates of development slow down and juvenile stages of ancestors become the adult features of descendants.) Many central features of our anatomy link us with fetal and juvenile stages of primates: small face, vaulted cranium and large brain in relation to body size, unrotated big toe, foramen magnum under the skull for correct orientation of the head in upright posture, primary distribution of hair on head, armpits, and pubic areas. ... In other mammals, exploration play and flexibility of behavior are qualities of juveniles, only rarely of adults. We retain not only the anatomical stamp of childhood, but its mental flexibility as well.

The proto-human young were nurtured for progressively longer periods of time as the size of the human brain decreased. The period of dependency began, of necessity, to exceed the four or five years on which the child was dependent upon breast feeding for sustenance. Because dependency of the young and the need to protect and feed them extended beyond the period of nursing, it is reasonable to assume that members of the kin network became engaged in part in this child care. This would be an evolutionarily efficient strategy, since kin fostering would also result in the passage of genetic material into the next generation. Neotenic development decreases the distinction between adult and child-like characteristics, blurring the distinctive traits that release affectionate care and thus extending the period of time and the kinds of people engaged in affectionate nurturing. One could hypothesize that the generalization of the nurturing response would also extend to disabled and sick members of the band, resulting in a higher survival rate from illness and accident.

When human beings began to rear other animals, perhaps by bringing home the young of adults killed in the hunt, they extended their opportunities for sensual involvement in nurturing activities. The care of animals became facilitated both by the practical value of the animals themselves, and by the pleasure and physiological rewards of caring for the animals. If nurturing plants and animals had some of the same rewards as caring for other human beings, then we would also expect the health of those groups who practiced domestication to have been improved both by better nutrition and by the direct beneficial effects of the increased opportunity to engage in nurturing activities.

Domestication of plants and animals also extended the opportunities for rearing human children. The limited resources available to hunting and gathering tribes requires them to space out childbearing. The increased food resources and opportunity for permanent settlement afforded by agriculture permits a greater frequency of childbirth (Fisher, 1986).

Agriculture was fully established some 10,000 years ago, providing for humankind a continuous and almost universal contact with and nurturing experience toward plants and animals. This engagement persisted throughout the history of civilization until the last 200 years. In those two centuries, only 10 to 15 generations, a trivial time in the genetic history of human beings, there has been an extraordinary disengagement on the part of people concerning the care of animals and plants. This trend began well before the industrial revolution, with its enclosure and expropriation of public lands, and its shift in agricultural practices to support trade in grain, wool, and cattle (Williams, 1973). Shifting patterns of agriculture, with the displacement of small or peasant farmers, continued into the beginning of this century in Europe, and continues of course in South and Central America to this day. People were not attracted to labor in the cities' mills, they were driven to it. Thomas Hardy, describing the depopulation of rural England, wryly remarked:

A depopulation was also going on. ... These families, who had formed the backbone of the village life in the past, who were the depositories of the village traditions, had to seek refuge in the large centers; the process humorously designated by statisticians as "the tendency of the rural population towards the large towns," being really the tendency of water to flow uphill when forced by machinery.

THOMAS HARDY, *Tess of the D'Urbervilles*
Penguin Edition, 1984, chapter 51

Since the industrial revolution, there has been an enormous shift of people into cities and away from any contact with the rearing of animals or the care of gardens and orchards. In the space of two centuries, the United States and western Europe went from a population that was only 10% urban to one that was 90% urban. In the United States, 1910 was the first year there were fewer farm workers than industrial laborers, and now farm labor makes up only a small fraction of the work force. Many of those remaining laborers have tasks such as seasonal harvesting that are divorced from the care of animals, although they do provide some contact with nature. In the relatively few years since industrialization there has been a radical transformation in the physical relationship between human beings and other living things, with a very large part of the population being excluded from contact with and care of living things other than their own children. In 1800 the world population was a little less than one billion. By the year 2000 the population of the world's five largest cities will total one billion people. In the past 50 years, the fall in birth rate, the rise in divorce rate, and the increased frequency of people who are electing to live alone has resulted in an increasingly large fraction of the population that has little or no experience in the care and nurturing of human animals.

Using the western or industrialized world as an example, never has there been a human population that has spent so little time in physical contact with animals and plants, and has devoted so little time to the nurturing of its own young or the care of animals. We have no idea what the cognitive, emotional, and physiological consequences of that change have been.

Part of our ignorance is willful ignorance. We are trying to convince everyone to have fewer children. If Americans are committed to motherhood and apple pie, the behavior scientists in this country are committed to the necessity of limiting population growth, and to the full inclusion of women in the society. Attainment of both of those goals would not be facilitated by a realization that nurturing other living things might have positive or beneficial health and emotional consequences. The belief that taking care of living things is necessary for health or for the attainment of full human potential is not a "serviceable" one. It does not justify the important social agendas of population control and universal employment.

Awareness of the emotional and health benefits of tending for living things does not imply that men and women in industrial nations must return to unlimited population growth. During the entire growth of civilization, human beings have satisfied their needs for nurturing activities by caring for plants and animals. The rewards of caring direct us toward a return to engagement with the living world.

The awareness of what taking care of animals can do for us occurs at a time when it is becoming clear how much we must be concerned with the preservation of the earth's living environment. It is now necessary to make a direct connection between our need to care for other living beings and the need to protect our fragile environment. In the profound reformation of our consciousness of nature, we are losing our long-held idea of nature as life outside of us; a powerful, potentially dangerous being, from which humankind must wrest the conditions of its existence. Nature is no longer outside of us, nature is no longer a powerful adversary that must be conquered, tamed, or endured. The industrial revolution supplied the tools both to conquer and to shrink the natural world, and nineteenth-century romanticism simultaneously glorified the wildness of nature and its conquest. In the twentieth century, the reduction of nature has been completed, and we have attempted to assuage our loss, and to recover what we have lost by the creation of a new metaphor for the link between our culture and nature. That metaphor does not create a new aesthetic, as the romantic image of nature once did; it has created a new ethic.

Nature is no longer a kind of demonic god that must be conquered or colonized for the glory or progress of men. Behemoth and Leviathan are endangered species. They have become tender infants in need of protection rather than terrible symbols of the demonic forces that can crush man's existence. One could speculate that if Captain Ahab were resurrected today, he would be driving his craft between the Japanese trawlers and the whale with the same reckless abandon with which he pursued the whale a century earlier. He would be protecting the whale as a tender child, rather than relentlessly attacking a brutal parent. When Ahab pursued the whale that maimed him, he saw, in nature, the author of the blind fury that shaped man's frail life. He saw nature as an all-powerful parent that both nurtured and killed its young. Now he would recognize that the blind fury in our lives is not a demonic nature, but humankind itself. Now nature must be protected from man, not man from nature.

We have conquered all of nature but ourselves. No part of the world is safe from our dangerous influence. The rain forests of Brazil and the Arctic ice all bear the scars of pollution, or of exploitation. We must now guard the ozone layer, the very outermost atmosphere of the earth's envelope, as carbon dioxide and fluorocarbons change the climate of the entire earth.

We now must care for all the living world. Wilderness is a collection of fragile species that must be rescued, nourished, and protected. There no longer is any true wilderness. Nowhere on the globe does life exist independent of human activity. Our growing interest in nourishing and caring for plants and pets in our private spaces is reflected in the growing knowledge that we must preserve and care for the life of the entire planet. Perhaps the work of the paleolithic cave painters of the Dordogne is nearing completion. They brought the first animals out of the world into the interior domain of human perceptual space. Now all of the world's animals and plants must be part of the human order of thought if they are to survive. We must continuously think and rethink the natural world if we are going to sustain it in its complexity and fullness.

What conclusions are to be drawn from this line of reasoning?

There is a critical need for continued and augmented exploration of the emotional and health value of nurturing living things. The investigation should include people at all phases of the life cycle. We should certainly continue to examine relationships with pets like dogs, cats, birds, small mammals, reptiles, aquarium fish, and horses, but should also observe people with gardens, farmers, and 4-H children with their animals, bird watchers, and wild bird feeders. The variety of human feeling and nurturing activities displayed in relationships with these diverse living things will permit an examination of all the emotional and behavioral factors contained in that global term, nurturing. We need to know as much about the effects of caring for living things as we are beginning to learn about exercise and touch.

We know already that there are immediate and favorable emotional and physiological changes proceeding from contact with pets, and, by inference, from the nurturing of domestic animals, gardens, and house plants. The active contemplation of the natural world also has the ability to reduce tension and to integrate the sense of self. This knowledge must be used to reinforce those movements in the society that are beginning to redress the radical deprivation of nurturing contact that has been accelerating over the past two centuries.

There has to be a linkage built for ourselves and our children between the care of pets and gardens and the care of the increasingly fragile natural environment (Bustad, 1986). We have to learn to benefit from nurturing our domestic animals, but we then have to learn to extract the same pleasure from nurturing at a greater distance, from caring for that part of the living environment that must be maintained as a thing in itself and not made into a cultural object.

Our zoological gardens should take a leading role in forming this new linkage between care of pets and domestic animals, and preservation of the natural world. Zoos must now, perforce, tame and breed animals that are threatened in the wild. Their curators must learn both to breed the animals and to preserve knowledge about their behavior so that eventually they can be released to recreated natural environments. Zoos, then, have a dual mission, that of providing a plea-

surable education about living nature, and of preserv-
ing endangered wildlife. The ethic inherent in those
goals should be underlined through a variety of education programs. Zoos should lend their expertise in a
cooperative venture with hobbyists to replace imported
birds, amphibians, reptiles, and fish with captive bred
animals. With the shortage of basic research funds, it is
probable that the captive breeding of many species will
be implemented by dedicated amateur naturalists. In
similar fashion, the zoo is in an ideal position to teach
people how to use their pets to gain a wider understanding of wild animals and of ecological systems. In
the most general sense, the zoo should be the starting
point for a new kind of science curriculum that is
oriented toward observation rather than experimentation, and that brings children back to an appreciation of
the appearance of the natural world. We need to know
how to identify trees and plants and birds, and to understand the organizing principle in the structure of
plants and animals. We need to know how to look at
the behavior of animals and at the operation of ecosystems. The educational activities of zoos should be the
start of a reform of our educational system so that
children can learn to see the natural world around
them. If Paul Shepard's (1978) intuitions are correct,
then learning plant and animal taxonomies should
greatly facilitate other kinds of learning. Science should
be taught by observation; children should learn a biology of the senses, an engaged science that lets them
both apprehend and care for the natural world.

Lastly, we must learn to use our ability to gain and
preserve ideas of natural things, to express our care for
the living world through active observation and
through attentive contemplation. It is a way of finding
solace and peace in that part of nature we must leave to
itself and visit with as little interference as possible.

By drawing healing from nature through a return to
the essentially human activities of care and contempla-

tion, we may be able to find the strength to restrain our own materialism, and so prevent the destruction of that vast structure of appearance that constitutes the natural world.

References

Baun, M., N. Bergstrom, N. Langston, and I. Thoma. 1984. Physiological Effects of Petting Dogs: Influences of Attachment. In *The Pet Connection: Its Influence on Our Health and Quality of Life*, eds. R. Anderson, B. Hart, and A. Hart, St. Paul, MN: Grove Publishing.

Bowlby, J. 1966. *Maternal Care and Mental Health*, pp. 30–45. New York: Schocken.

Bustad, L. In press. *The Living Bond: An Afterward, Learning and Living Together—Building the Human/Animal Bond*. The Delta Society.

Cain, A. 1983. A Study of Pets. In *New Perspectives on Our Lives with Companion Animals*, eds. A. Katcher and A. Beck. Philadelphia: Univ. of Pennsylvania Press.

Eddy, J., L. Hart, and R. Boltz. 1986. *Service Dogs and Social Acknowledgement of People in Wheelchairs: An Observational Study*. Paper presented to the Delta Society International Conference.

Fisher, H. 1986. *The Human Divorce Pattern in Cultural Perspective*. Paper presented at the Man and Beast Revisited Symposium, Smithsonian Institute.

Friedmann, E., A. Katcher, J. Lynch, and S. Thomas. 1980. *Animal Companionship and One-Year Survival of Patients after Discharge from a Coronary Care Unit*. Public Health Reports 95:307–12.

Friedmann, E., B. Z. Locker, and S. A. Thomas. 1986. *Effect of a Pet on Cardiovascular Responses during Communication by Coronary Prone Individuals*. Paper presented to the Delta Society International Conference, 1986. Abstract p. 137.

Gould, S. 1981. *The Mismeasure of Man*. New York: Norton.

Grossberg, J., and E. Alf. Interaction with Pet Dogs: Effects on Human Cardiovascular Response. In *The Journal of the Delta Society* 2:20–7.

Hardy, T. *Tess of the D'Urbervilles.* New York: Penguin Books, 1984.

Harlow, H., and M. K. Harlow. 1962. Social Deprivation in Monkeys. In *Scientific American* 207:136–44.

Jenkins, J. 1985. Physiological Effects of Petting a Companion Animal. In *The Journal of the Delta Society* 2:60 (abstract).

Katcher, A. 1981. Interactions between People and Their Pets: Form and Function. In *Interrelations Between People and Pets,* ed. B. Fogle. Springfield, IL: Charles C. Thomas.

Katcher, A. 1985. Physiologic and Behavioral Responses to Companion Animals. In *The Veterinary Clinics of North America: The Human-Companion Animal Bond,* eds. J. Quackenbush and V. Voith, 15–2:403–10.

Katcher, A., and A. Beck. 1986. Dialogue with Animals. In *Transactions and Studies. College of Physicians of Philadelphia* 8–2:105–12.

Katcher, A., and E. Friedmann. 1980. Potential Health Value of Pet Ownership. In *Compendium on Continuing Education for the Small Animal Practitioner* 2:887–91.

Lockwood, R. 1983. The Influence of Animals on Social Perception. In *New Perspectives on Our Lives with Companion Animals,* eds. A. Katcher and A. Beck. Philadelphia: Univ. of Pennsylvania Press.

Lynch, J. 1977. *The Broken Heart: The Medical Consequences of Loneliness.* New York: Basic Books.

Messent, P. 1983. Social Facilitation of Contact with Other People by Pet Dogs. In *New Perspectives on Our Lives with Companion Animals,* eds. A. Katcher and A. Beck. Philadelphia: Univ. of Pennsylvania Press.

Ory, M., and E. Goldberg. 1983. Pet Possession and Life Satisfaction. In *New Perspectives on Our Lives with Companion Animals,* eds. A. Katcher and A. Beck. Philadelphia: Univ. of Pennsylvania Press.

Ruby, J. 1983. Images of the Family: The Symbolic Implications of Animal Photography. In *New Perspectives on Our Lives with Companion Animals,* eds. A. Katcher and A. Beck. Philadelphia: Univ. of Pennsylvania Press.

Shepard, P. 1978. *Thinking Animals: Animals and the Development of Human Intelligence.* New York: The Viking Press.

Spitz, R. A. 1945. Hospitalism: An Inquiry into the Genesis of Psychiatric Conditions in Early Childhood. In *The Psychoanalytic Study of the Child* 1:53.

Tuan, Y. 1984. *Dominance and Affection: The Making of Pets.* New Haven: Yale Univ. Press.

Williams, R. 1973. *The Country and the City.* New York: Oxford Univ. Press.

Attitudes toward Animals: Origins and Diversity

Harold A. Herzog, Jr.
Gordon M. Burghardt

While Katcher and Beck are well-recognized figures in human-animal–bond research circles, Hal Herzog and Gordon Burghardt are not. Yet Herzog and Burghardt have published several provocative papers on the interaction between humans and animals, with a special emphasis on human attitudes toward animals and the underlying psychology of the contradictions and paradoxes in human behavior involving animals. They comment that it would be difficult to overestimate the significance of animals in the social and psychological life of our species. Images of animals are everywhere: in our language, religions, dreams, television programs, and folklore. Responses to animals range from disgust to reverence and from love for a pet to the paralyzing fear of phobic experiences. Herzog and Burghardt examine a wide range of human-animal interactions and show the incredible richness of human experience with animals. We have only just begun to tap these topics—Herzog and Burghardt provide a stimulating introduction. EDITOR

It would be difficult to overestimate the significance of animals in the social and psychological life of our species. Images of animals are everywhere: in our language, religions, dreams, television programs, and folklore. The feelings that we exhibit toward our fellow creatures are intense, complex, and paradoxical. Responses to animals range from the disgust we feel when confronted with a bloated tick to the reverence for animals as deities in so-called primitive cultures; from the love of a child for a pet bunny to the paralyzing fear of phobic experiences when confronted by a harmless spider.

In recent years there has been increasing interest in human-animal relationships by investigators from a variety of disciplines. We will not attempt a synthesis of the growing literature on attitudes toward animals, but will follow a different course. For the past decade we have been exploring the diversity and origins of human-animal relationships, and our research has taken us into some rather odd places: cockfights in the United States and Latin America, slaughterhouses, and most recently, the world of supermarket check-out-counter magazines. In this article, we will summarize some of our findings and speculations that bear on the subject of attitudes toward animals. We will also briefly examine alternative methods of gathering information that do justice to the richness of human experience with animals.

Origins: Evolution and Biology

Clearly such factors as age, education, urbanization, and social class influence the formation of attitudes. But are attitudes toward animals simply culturally-determined practices and emotions passed on through generations via language and observation? The great diversity across cultures in attitudes toward animals often leads to emphasis on cultural influences and learning. For example, as Serpell (1986) describes in an excellent overview of cross-cultural and historical differences in human-animal relations, in different societies dogs may be revered or relegated to the stew pot. But let us entertain the idea that such cultural variation dances upon a deep, genetically shaped reservoir of universal predilections and emotions. Perhaps behind the apparent diversity of attitudes are some underlying biological processes that are the result of natural selection.

The model suggested here parallels recent findings on human facial expressions. There is significant cultural variation in emotional expression, and also a repertoire of components that appear to be innate and universal (Ekman, Friesen, and Ellsworth, 1972). Kellert (1983) has argued that many of our responses to animals derive from our hunter-gatherer heritage. Is this heritage not genetic as well as cultural?

We suggest that human attitudes toward animals have some connection to natural selection. The evidence is of two types: 1) evidence of innate responses or the readiness quickly to learn certain things and 2) plausible scenarios indicating that the value-related considerations *could* have been produced or influenced by natural selection. There is space to discuss briefly only a few of these factors. We can classify these responses or attitudes into those that have been shaped directly or indirectly by selective pressures. By *direct* we mean aspects of our responses that are derived from human evolutionary history with animals. By *indirect* we mean attitudes to animals that are anthropomorphic generalizations of responses that originally evolved toward other people. Mixed effects could also occur in which the original selective forces cannot be distinguished.

Direct Selective Effects

a) Fear of snakes: Whether the fear of snakes (*ophidiophobia*) is learned or innate has been debated for many years. What is clear, however, is that many species of non-human primates express marked emotional reactions to live snakes or elongated models even when they have had little or no previous experience with them (Mundkur, 1983). Fear of snakes is also one of the most common human phobias. Why should we and other primates be afraid of snakes? Possibly because some are poisonous and deadly. Before guidebooks were available, hesitating to identify a snake might have been unwise; indeed, you could have been selected out. In Asia and Africa, where early humans evolved, there are no simple rules for quickly discriminating poisonous from non-poisonous species as there are in the United States and Europe. Thus we see no problem in ascribing the fear of snakes (and also spiders, of which a few are deadly) to genetic factors. The consequences of curiosity could kill more than a cat.

b) Domestication: At certain points in human evolutionary history, there might have been selection favoring individuals who exploited, tamed, and effectively cared for domestic stock. Today research is showing that farm animals given individual attention and perceptive care are more productive and healthier than those given automated, impersonalized care. Could not the extraordinary care and devotion many individuals lavish on domestic animals be derived, not from seeing them as child-substitutes and members of the family, but from selection of human characteristics that predispose to successful animal husbandry? This would be particularly true in societies where animal ownership was equated with status and wealth. In a similar vein Voith (1985) has put forth the intriguing argument that the presence of animals is soothing to people because in our evolutionary past, the proximity of non-frightened animals signaled that there were no nearby predators.

c) Rarity and population distribution: We tend to favor rare animals. This may seem somewhat arbitrary insofar as the suffering of an individual animal is concerned, but the death of a cow is not the front-page story that a beached whale merits. While knowing a species is endangered is intellectually transmitted knowledge, our concern for them, often so emotional, may be derived from the value put upon rare animals in early human history. Rare animals or their body parts played important roles in rituals, costumes, medicine, and social status. (Unfortunately, this is still the case in some parts of the world as evidenced by the demand for rhinoceros horn on the Arabian peninsula as dagger handles and in the Orient as aphrodisiacs.) Some species may be very plentiful in a small restricted area, say an island, but still be vulnerable to extinction and over-exploitation. Our concern for animals found only in restricted areas may be related to the needs and experiences of hunter-gatherer societies and have a biological component (Wilson, 1984). Of course, it is quite apparent that humans have caused the extinction of many local populations and species; the story is undoubtedly a complex one. But the fervor and intensity with which small segments of our population fight to save threatened species, albeit with an emphasis on large mammals and aesthetically pleasing species (Burghardt and Herzog, 1980) needs explanation from more than a cultural perspective.

Indirect Selective Effects

a) Baby releasers: The attractiveness of juvenile animals has long been recognized, although it took Konrad Lorenz to isolate the specific features of infants (large eyes, large cranial expanse, short stubby limbs) that are the basis for our attraction to kittens, puppies, and other infant animals. Our response to "cute" animals is most surely based on the generalization across species of parental-care responses that were essential to the survival of young in species with prolonged juvenile dependency, such as humans.

b) Juvenile behavior: Some aspects of the behavior of juvenile mammals are as tugging as their appearance. One of us (GMB) remembers being enchanted by the awkward movements of a newborn moose at the Milwaukee Zoo. Though a moose even when young is not

your standard cute animal, its behavior made it irre-
sistible. As with baby releasers, our attraction to these
attributes of animals could very well be based on
genetic predispositions to attend and respond to juvenile humans. Slightly awkward children, lacking in
seriousness, need parental safeguarding and attention.
The term "toddler" describes this stage well.

c) Communication method: Our sensory abilities make
us more aware of the cries of puppies than the high-
frequency cries of baby bats or the pheromones given
off by frightened mice. But in our evolutionary history,
sensory abilities evolved to deal with *our* problems, not
other animals'. Thus, we are biased toward those species with which we can at least have the illusion we are
communicating or at least recognize their signals for
what they are.

d) Pain and suffering: The issue of pain and suffering
by animals is perhaps the most frequently discussed
and controversial attribute of them all. We merely point
out that we are often most adamant and emotional
about the suffering of animals that respond as we think
we might in similar circumstances. Thus, we are much
more likely to show moral outrage to mistreatment of
species that show physiognomic or psychological similarities to ourselves (Burghardt and Herzog, 1980).
Many are upset by cruelty to primates; few evidence
similar sensibilities toward brutality to reptiles. We ignore potential suffering in species to which we have difficulty relating. Such occasions include boiling of live
lobsters, spraying cockroaches, or hooking fish.

In short, we believe that many of our attitudes
toward animals may be based on biological, genetic predispositions. The source of these may be in anthropomorphic generalizations of responses that evolved
toward conspecifics, as in our attraction to big-eyed
baby harp seals, or neotenic breeds of dogs such as Chi-
huahuas. Other responses may be the product of increased fitness that accrued to individuals who treated

animals in specific ways—for example, avoidance of snakes. The notion that our cognitive evaluations of animals and their treatment is superimposed on more primitive affective responses that were adaptive many thousands of years ago has implications for any alteration of our behavior and attitudes toward animals.

The Influence of Experience on Attitudes

While it is easy and fun to speculate as to the biological roots of our feelings about animals, it is obvious that personal experiences have greater effects on individual differences in attitudes; whether we are "dog people," "cat people," or what we might call "animal misanthropes" (Brown, 1985). Consider, for example, the animals we eat. As populations have moved from rural to urban settings, dominant cultural attitudes toward animals and their use as a source of protein have made concomitant changes. As fewer individuals are raised around livestock, we become psychologically distanced from the fact that the convenient packages of pork, beef, and chicken that we purchased at the supermarket were once living creatures that were bred, raised, and finally killed for our benefit. Except for hunters and slaughterhouse workers, few in Western cultures have the experience of being involved with the slaughter of large animals. We might expect that this experience would have a significant effect on one's attitudes toward the consumption of animals. Several years ago, one of us (HAH) had the opportunity to investigate the psychological effects of the experience of slaughtering livestock on a group of relatively naive students working on a college farm crew (Herzog and McGee, 1983). Warren Wilson College is a small, liberal arts college located in the mountains of western North Carolina. A unique feature of the college is the work program in which all residential students work 15 hours per week in exchange for room and board. The college maintains a farm that produces beef cattle and hogs, and students are involved with all aspects of animal production including slaughter.

In order to describe changes in the attitudes of students toward animals as a result of the experience of slaughtering, the members of the crew who had participated in slaughtering were given questionnaires and subsequently interviewed. Surprisingly, the students overwhelmingly, with only a couple of exceptions, felt that slaughtering had been a valuable and significant experience in their lives. Reasons they gave varied from the mundane ("I now know about the various cuts of meat so I can shop more effectively.") to the more profound, in which students described their experience with the death of an animal in metaphysical terms. Students who were relatively experienced at slaughtering were more likely than novices to admit that they sometimes got a feeling of accomplishment from the experience, while inexperienced students were more likely to say that they avoided eating meat after a slaughtering session.

We also gave the students a scale designed to measure their general feelings about the use of animals for medical research, sport hunting, rodeos, and other animal-using activities. On the basis of their responses students were grouped into "doves" and "hawks." The results indicated that "hawkish" students were more likely to admit that slaughter gave them a feeling of accomplishment, while the "doves" were more likely to say that they sometimes felt guilty after slaughtering.

In general, the students felt that their farm crew experience, which included slaughtering as well as raising animals, had helped them clarify their attitudes about eating other creatures. They felt they were special. They knew things others do not know—such as where meat comes from and what a warm carcass feels like. They had, at least once in their lives, assumed the moral responsibility of carnivory, and they acknowledged that this experience had influenced their attitudes toward animals.

Individual and Subcultural Diversity of Attitudes

Last summer, a man who seemed to be in a bit of a panic phoned our laboratory at the University of Tennessee where we study the behavior of reptiles. For the past ten years, he and his wife had kept a pair of large boa constrictors caged in their suburban living room as pets. That morning the female had, with no warning, produced a litter of 35 baby boas. The man was frantic. He had no idea what to do with the snakes and wondered if we could help him with temporary housing. When we mentioned that he might make a considerable sum of money, several thousand dollars, by selling the offspring to pet shops he balked, stating with all sincerity that finding a good home for all of the babies was his paramount concern, rather than financial gain. In addition, he said that he and his wife did not have any children, and the two adult boa constrictors were "just like members of the family."

This story illustrates what we might term "idiosyncratic diversity" of attitudes toward animals. While most people view snakes with something akin to revulsion, this quiet, apparently conventional middle-class couple had two large boa constrictors as child-surrogates—a niche more typically occupied by dogs or cats. While such individual differences in attraction and revulsion directed toward different species are fascinating, they are not always easy to explain (Kidd and Kidd, 1987).

In addition to individual differences in attitudes toward animal species and their use, there are also cultural norms that affect our feelings about animals and what constitutes acceptable treatment of them. And, of course, within a large and diverse country such as the United States, there are numerous subcultures that may have quite divergent views about the treatment of animals. As an example, we will use a group that is surprisingly common in our area of the southern Appalachian mountains—cockfighters.

Many people find the subject of "blood sports" distasteful, to say the least, and it might be argued that it is inappropriate even to discuss the topic in a volume on human- animal relationships. But, simply ignoring a topic that we might not like does not further the scientific investigations of human-animal interactions. We must be willing to explore "the good, the bad, and the ugly." Thus, it is essential that those of us interested in human-animal relations as a broadly conceived area of research attempt to understand the factors that underlie the reasons why such an activity is repugnant to most people but is held by a dedicated minority to be man's noblest sport. In addition, cockfighting, like drug use, will only be eradicated when individuals cease to be absorbed by it. Thus, an understanding of the activity is especially critical to those who would like to see such activities eliminated.

Visitors to the Southern Appalachian mountains casually driving through the countryside may be puzzled occasionally to see fields containing wire cages, each housing a single rooster. Discreet inquiries will reveal that these are not ordinary barnyard fowl but gamecocks, animals maintained solely for the clandestine sport of cockfighting. Several years ago, we became acquainted with a number of "cockers" as they are called and began to investigate a widespread, though generally illegal, activity throughout much of the United States. We were surprised to discover that in the United States there are three magazines (*Grit and Steel, The Gamecock,* and *Feathered Warrior*) exclusively devoted to cockfighting, each of which contains ads for equipment and paraphernalia, announcements of upcoming fights, and advice columns. We were not aware that there was a formal lobby group, the United Gamefowl Breeders

Association, whose goal is to combat the growing number of attempts in state legislatures to strengthen anti-cockfighting statutes. Nor did we suspect that, just as there is a football season, there is a cockfighting season (Thanksgiving to the Fourth of July). Over a period of several years, we interviewed many rooster-fighting aficionados. We found a diverse group of individuals, young and old, almost exclusively male, who, for the most part, were intensely proud of their involvement with cockfighting despite the fact that their sport is viewed rather dimly by the majority of Americans. Many were socialized into cockfighting at an early age. It is not unusual for young kids to be brought to fights by their fathers. Generally the cockers were not at all secretive about their interest in cockfighting and were quite willing to discuss various aspects of their sport.

While a detailed description of the culture of cockfighting is beyond the scope of this chapter, we would like to mention some of its more salient aspects. (The interested reader may want to consult sources such as Pridgen [1938], Rupport [1949], Del Sesto [1975], or Herzog [1985] for more detailed information.) Game roosters are strains of chicken bred exclusively for combat. The fighting strains have been kept reproductively separate from production strains for many generations. Strains have names such as Madigan Gray, Roundhead, Butcher, and Claret and vary widely in color. Owners typically take great pride in the appearance of their animals and are vehement on the subject of their care. In fact, cockers commonly defend their sport on the grounds that their roosters are much better cared for than are factory-farm broilers or even most barnyard fowl.

Fights are highly organized affairs that take place in specially built arenas complete with refreshment stands and public address systems. An intricate set of rules governs the conduct of the fight. During the actual fight, there is a referee present in the pit as well as the "handlers" of the cocks. Gambling is an intrinsic part of the enterprise, and spectators yell out bets and odds throughout the fight. Prior to each match, steel "gaffs" are attached to the stubs of the roosters' natural spurs. This, of course, means that cockfights are lethal affairs, and they end with the death of the loser in over 85% of matches (Herzog, 1978).

It is not, however, the details of animal husbandry and rules of the fight that are of interest here but the underlying attitudes that make cockfighting morally acceptable to a subgroup within American culture. It seems clear that cockfighters do not exhibit obvious identifiable psychopathology. Clifton Bryant and the late William Capel (unpublished manuscript) gave a variety of psychological tests to cockers and essentially found that they are quite typical of the rural population from which they are primarily drawn. They are no more psychotic or sadistic than any other group of predominantly lower-middle-class, rural males. In fact Capel (pers. comm.) found that cockfighters were psychologically indistinguishable from stock-car-race fans—the most popular Southern sport.

However, as with other groups of individuals with utilitarian approaches toward animals, such as hunters, ranchers, and medical researchers, cockfighters possess a set of values and attitudes that enables them to justify their sport (see also McCaghy and Neal, 1974). We have heard the following arguments from cockfighters:

1. Critics of cockfighting have never been to a fight and really do not understand the sport.
2. Gamecocks were meant to fight and do so naturally; therefore, it is unnatural not to pit them against each other. There are several variations of this theme. One variation is that God put gamefowl on Earth to fight; another is that humans have bred roosters to fight

and, thus, they have a right to enjoy cockfighting as a sport.

3. The bravery of the game rooster is a superb model for humans to emulate.
4. It is a family sport that keeps the kids off the street, as they are often involved in the day-to-day maintenance of the animals as well as in the exercise regime that goes along with preparation for upcoming fights.
5. Chickens are too dumb to feel pain.
6. Some of our great American heroes such as Andrew Jackson and Abraham Lincoln have been avid cockfighters.

Paradoxically, most of the cockfighters that we are acquainted with draw a clear moral distinction between cockfighting and other bloodsports such as dogfighting and bearbaiting, which also occur in the southern mountains (though much more surreptitiously). The majority feel that dogfighting is cruel and want nothing to do with it.

It is difficult to describe the attitudes and feelings that cockers have toward their roosters. It is clear that they have immense respect for their birds and truly admire that combination of courage, tenacity, and confidence that they refer to as "gameness." They take pride in the way their birds look, and of course, in the way they fight. For many, the attraction of cockfighting is really the challenge of breeding a better fighting bird, and some have a sophisticated understanding of genetics and selective breeding.

On the other hand, gamecocks are not pets in any sense of the word. They are animals to be used. The attitudes of cockers toward their roosters is clearly what Kellert (1980) has labeled "utilitarian." Gamecocks are of value only as long as they win. Tears are not shed over a dead loser. Animals that flee in the pit are an embarrassment to their handlers and breeders.

Finally, what draws individuals to what has been referred to as a deviant sport? The attraction is not sadism. Cockfighters do not get their thrills from watching chickens bleed to death. Rather, they seem to be drawn by their intense sense of competition and tradition, the excitement of betting, and the aesthetics of the fight.

In addition, cockfighters in the southern mountains are, for the most part, not violating the mores of the cultures in which they live (Herzog, 1988). They often make little attempt to hide their involvement in an illegal bloodsport. Indeed, they often publicly proclaim their enthusiasm by wearing cockfighting belt buckles or by proudly displaying on their cars "Sport of Kings" license plates complete with crossed gaffs. The fact that the football team of the University of South Carolina is officially designated the "Fighting Gamecocks" says something about the defacto acceptance of cockfighting as a Southern tradition.

Views of Animals in Popular Culture

Cockfighters represent an unusual subculture with ideas about the uses of animals that are at odds with the majority of American society. But, there is also wide variation of attitudes toward animals within what is sometimes termed the "dominant culture." These differences in value orientations have been extensively documented by Kellert (1980), who has developed sophisticated survey techniques to assess and compare attitudes about animals.

However, there are limits to the types of information that can be obtained through surveys, and we need to be vigilant for alternative windows into the nuances of human-animal relations. Some years ago, Clifton Bryant (1979) pointed out that images of animals are ubiquitous in American language and culture, and we have recently begun to use the portrayal of animals in popular media as a means of assessing aspects of attitudes toward animals that are not easily uncovered by more traditional methods. Specifically, we have found that the "supermarket tabloids," periodicals such as *The National Enquirer*, offer an unusual, and sometimes humorous, glimpse into the roles of animals in our psychological life. Take, for example, the depiction of dogs in the tabloid press.

Our canine friends have a multitude of roles in these papers, as they do in our lives. The bulk of dog stories and pictures portray the dog in its most common role, that of companion. These items most often take the form of images of puppies kissing children, or cuddling equally cute non-canids such as kittens, bunnies, or baby squirrels. Similarly, there are numerous stories that portray unusually intense bonds between master and dog: "John Paul Getty Treats a Dog Better Than His Dying Boy" (*National Enquirer* 4/1/86); "Widow Marries Dog To Get Tycoon Hubby's Fortune" (*World Weekly News* 1/6/87).

There is another category in which the dog is stereotypically man's best friend. The motif in these stories is that of the brave and loyal pooch saving the master's life. The image of the dog as savior is one of the most pervasive images of animals in popular culture and runs from Appalachian folk tales through TV shows such as "Lassie" and "Rin Tin Tin." The tabloid press is no exception as witnessed by headlines such as "Loyal Dog Saves the Life of Freezing 80-Year-Old Man" (*National Examiner* 4/15/86).

But, dogs are also capable of the ultimate betrayal; they can eat your children. While the world of sensationalistic papers is populated with hostile beasts—man-eating tigers, crocodiles, vampire bats, and giant snakes—it is the dog that accounts for the bulk of stories where humans are attacked by animals. Surprisingly, the portrayal of our canine companions as potential predators is based on fact. In the Western world, dogs are the animals (aside from allergic reactions to insects stings) most likely to represent a real threat to human life and limb (Borchelt, Lockwood, Beck, and Voith, 1983). This is amply demonstrated in the tabloids with such exaggerated headlines as "St. Bernard Bites off Mailman's Leg ... He Buries It in Backyard" (*Sun* 10/14/86) and "Heavy Metal Rock Music Turns Poodle into Vicious Killer" (*Sun* 3/24/87).

But humans can also treat animals cruelly, as evidenced by headlines such as "Horrified Mom Learns Her Neighbor Is Butchering Little Dogs to Make ... Puppy Chow" (*World Weekly News* 6/3/86) and "Breeders Kill Dalmation Puppies ... Spots Were in the Wrong Place" (*Sun* 10/26/86).

There are several other categories of stories in which dogs are depicted. Many dog pictures and stories have an anthropomorphic slant in which the animals are dressed in human attire or engage in human activities "Why Drunken Poodle Had To Give Up Booze," (*National Examiner* 5/6/86). Another category includes dogs with supernatural powers or experiences: "Dog Is Reincarnation of 4-Year-Old Boy," (*World Weekly News* 4/21/85); "Ghost of Dog Returns to Beloved Master" (*National Examiner* 5/14/87). Finally, the role of dogs in human life is reflected in the numerous cartoons in which dogs are the central figure. In the bulk of cartoons, the punch line is that the dog is smarter than the owner.

Our thesis is that the themes underlying the bizarre and ridiculous tabloid stories reflect the sometimes contradictory roles that animals play in our lives as well as the psychological factors that mediate our feelings toward animals. For example, the headline "Vet Slapped With $100,000 Suit for Sparing Pooch's Life ... 'How Do You Kill a Dog Who's Wagging His Tail and Licking Your Face?'" (*World Weekly News* 5/20/86) says something fundamental about human nature. Whether or not these stories are true is irrelevant; these tales are, in a sense, metaphors illustrating archetypes of human feelings and beliefs about animals that are not uncovered by more traditional measures of attitudes.

Animals and Human Experience

Finally, we would suggest an additional source of information about our psychological responses to animals—human consciousness. After having been banished from psychology with the ascendancy of behaviorism, the concept of consciousness as a topic of scientific scrutiny, even in animals, is making a comeback (see Burghardt, 1985; Griffin, 1981; Pollio, 1982). Though the traditional problems remain (such as how to get reliable data), in recent years a number of psychologists have concluded that there are some important questions that can only be investigated by studying subjective reports of conscious experience, a position referred to as phenomenology. We concur and suggest that methods that phenomenologists use to investigate "being in the world" might be fruitfully applied to the study of human-animal relationships.

The sometimes mystical relations between horse and rider, pet and owner, and even hunter and prey have eluded scientific description. Poets or fiction writers are, perhaps, more likely to capture these experiences than scientists. Thus the prose of Norman McClean in his classic story "A River Runs Through It" (1976) offers great insight into the psychology of fishing; insight currently beyond any scientific analysis:

> *The body and spirit suffer no more sudden visitation than that of losing a big fish, since, after all, there must be some slight transition between life and death. But, with a big fish, one moment the world is nuclear and the next it has disappeared. That's all. It has gone. The fish has gone and you are extinct, except for four and a half ounces of stick to which is tied some line and a semitransparent thread of catgut to which is tied a little curved*

piece of Swedish steel to which is tied a part of a feather from a chicken's neck. ...

Poets talk about "spots of time," but it is really fishermen who experience eternity compressed into a moment. No one can tell what a spot of time is until suddenly the whole world is a fish and the fish is gone. I shall remember that son of a bitch forever.

While intrinsically less rigorous than techniques such as attitude surveys, the phenomenological methods of loosely structured interviews and analyses of text have the potential of providing insight into the meaning of animals in human life that could not be gained through the more traditional methods of behavioral scientists.

This is ultimately a matter of levels, for we can go from physiology to genetics to individual learning to cultural settings and habitat. Methodologically we can observe, experiment, and study artifacts, opinions, and political systems. But, in the end, we cannot ignore the deep-seated experiential values concerning animals held by every person, for these provide the ultimate source of both our personal insight and scientific understanding.

Summary

In this article, we explore a variety of topics that illustrate the sources and diversity of attitudes that individuals exhibit toward animals. We have suggested that we must look to human evolutionary history as well as to individual experiments for insight into the muddled reasons as to why we treat animals as we do. The studies of inexperienced college students who slaughter cattle and hogs show that experiences with animal death can have a significant impact on attitudes about their use.

We have also examined the extraordinary diversity of feelings that individuals have about animals. Some of this variation is idiosyncratic; we have no idea why a middle-class couple from east Tennessee would prefer a pair of boa constrictors to a couple of dogs as pets. However, as southern Appalachian cockfighters demonstrate, some variation in attitudes is due to subcultural and regional differences regarding morally acceptable behavior toward animals.

Finally, we would like to emphasize that there are innumerable reservoirs of information, including such diverse sources as popular culture and human consciousness, that can be fruitfully tapped by investigators exploring the complexities of our relations and feelings about animals. We have only begun to scratch the surface of a topic that is developing into an exciting interdisciplinary field of study.

References

Borchelt, P., R. Lockwood, A. Beck, and V. Voith. 1983. Dog Attack Involving Predation on Humans. In *New Perspectives on Our Lives with Companion Animals*, eds. A. Katcher and A. Beck. Philadelphia: Univ. of Pennsylvania Press.

Brown, D. 1985. Cultural Attitudes Towards Pets. *Veterinary Clinics of North America 15*: 311–17.

Bryant, C. 1979. The Zoological Connection: Animal-Related Human Behavior. *Social Forces 58*: 399–421.

Burghardt, G. 1985. Animal Awareness: Current Perceptions and Historical Perspective. *American Psychologyist 40*: 905–19.

Burghardt, G., and H. Herzog 1980. Beyond Conspecifics: Is Brer Rabbit Our Brother? *BioScience 30*: 763–68.

Del Sesto, S. 1975. Roles, Rules, and Organization: A Descriptive Account of Cockfighting in Rural Louisiana. *Southern Folklore Quarterly 39*: 1–14.

Ekman, P., W. Friesen, and P. Ellsworth. 1972. *Emotion in the Human Face.* New York: Pergamon Press.

Griffin, D. 1981. *The Question of Animal Awareness: Evolutionary Continuity of Mental Experience*, 2nd ed. New York: Rockefeller Univ. Press.

Herzog, H. 1978. Immobility in Intra-Specific Encounters: Cockfights and the Evolution of "Animal Hypnosis." *Psychological Record* 28: 543–48.

Herzog, H. 1985. Cockfighting in Southern Appalachia. *Appalachian Journal* 12: 114–26.

Herzog, H. 1988. Cockfighting and Violence in the South. In *The Encyclopedia of Southern Culture,* ed. W. Ferris. Chapel Hill, NC: Univ. of North Carolina Press, in press.

Herzog, H. and S. McGee. 1983. Psychological Aspects of Slaughter: Reactions of College Students to Killing and Butchering Cattle and Hogs. *International Journal for the Study of Animal Problems* 4: 124–32.

Kellert, S. 1980. American Attitudes Toward and Knowledge of Animals: An Update. *International Journal for the Study of Animal Problems* 1: 87–119.

Kellert, S. 1983. How Animals are Perceived in America. Paper delivered at conference, Perception of Animals in American Culture, Washington, DC.

Kidd, A., and R. Kidd. 1987. Seeking a Theory of the Human/Companion Animal Bond. *Anthrozoös 1:* 140–57.

Maclean, N. 1976. *A River Runs Through It and Other Stories.* Chicago: Univ. of Chicago Press.

McCaghy, C., and A. Neal. 1974. The Fraternity of Cockfighters: Ethical Embellishments of an Illegal Sport. *Journal of Popular Culture 8*: 557–69.

Mundkur, B. 1983. *The Cult of the Serpent.* Albany, NY: SUNY Press.

Pollio, H. 1982. *Behavior and Existence.* Monterey, CA: Brooks/Cole.

Pridgen, T. 1938. *Courage: The Story of Modern Cockfighting.* Boston: Little, Brown.

Rupport, A. 1949. *The Art of Cockfighting: A Handbook for Beginners and Old Timers.* New York: Devin-Adair.

Serpell, J. 1986. *In the Company of Animals.* Oxford: Basil Blackwell.

Voith, V. 1985. Attachment of People to Companion Animals. *Veterinary Clinics of North America 15*: 289–95.

Wilson, E. O. 1984. *Biophilia.* Cambridge, MA: Harvard Univ. Press.

Horses in Society

Elizabeth A. Lawrence

Elizabeth Lawrence has spent a great deal of time in the last ten years examining different human-horse cultures and teasing out the essential threads of these interactions. The result of her research has been two excellent books on this topic with the first (Rodeo) winning the prestigious James Mooney Award for anthropological research. The chapter here is but a brief introduction to this topic. We learn of the extraordinary rapidity with which a people can embrace horses and take on the typical qualities of a mounted society —namely, pride, superiority and independence. The cowboy of the American West is a typical example. Whereas cowboys were no more than hired hands for the ranchers they served, they developed an imperious and arrogant attitude that was completely out of keeping with their station in life. Throughout history, horses have been perceived as symbols of victory and authority but they also have a more subtle message. Mounted police are viewed much more favorably than their motorized counterparts. Horses are associated with time and their measured tread and rhythmic gaits are comforting and healing to people caught up in the modern rat race. It is, for example, striking that there should be more horses in America today than at any time since the cavalry was disbanded. And yet, horses are not always seen as benign. Western ranchers are waging an all-out war to eliminate or greatly reduce the relatively small herds of wild mustangs. No one reading this chapter can doubt that the horse carries a powerful symbolic message and that horses provide human beings with exercise, beauty, grace, and a communion with nature that seems to restore the human spirit.
EDITOR

"Wherever man has left his footprint in the long ascent from barbarism to civilization we will find the hoof-print of the horse beside it" (Trippett, 1974, p. 9). Without the horse, the history and culture of mankind would have been far different. The status of the equine animals in human society is unlike that of most other domesticated species. Although humankind's first use of the horse probably was for food (Clutton-Brock, 1981, p. 80), the exploitation of its tremendous potential for power became the hallmark of many important early civilizations. With the harnessing of its strength and swiftness to provide mobility, the horse trans-formed human existence. Once the species had estab-lished a close relationship of interdependence with people, many equestrian societies spurned the meat of their working partners, typically establishing a strong taboo against horseflesh. The horse is undoubtedly the only animal whose consumption by Christians was specifically prohibited by papal decree (Harris, 1985).

That horses have been almost universally held in high regard is not difficult to understand, for "learning to control such a remarkable beast was probably the most exciting development in man's history, next to the invention of the wheel" (Hope and Jackson, 1973). It is generally believed that horses were first used to draw chariots in the ancient world before their widespread use for riding (Zeuner, 1963, p. 337), although some his-torians disagree. One of the earliest-known depictions of a person riding a horse is a painted wooden statuette from Egypt, dating to about 1800–2500 B.C. (Vesey-Fitzgerald, 1947). Archaeological data suggest that other species, such as the ox or onager, may have been ridden before horses (Downs, 1961; Zeuner, 1963, p. 315).

Evidence indicates that the domestication of horses had definitely taken place by about 3000 to 2500 B.C. (Zeuner, 1963, p. 337). Although it is generally accepted that the horse was not domesticated by Palaeolithic or Mesolithic people, certain intriguing discoveries have led some scholars to believe that the animal was being used at a much earlier date. An Ice Age cave painting from France, for example, shows a horse with a line across its head, which is strongly suggestive of a harness. Even more convincing is a 14,000 to 15,000-year-old horse-head carving from a cave site in the Pyrenees, which bears engraved lines resembling a halter that were added after the rest of the art work was completed. Additionally, the finding of horse teeth dating from 30,000 years ago, which show definite signs of the kind of wear produced only by crib-biting, provides another bit of tantalizing archaeological data. Since the habitual behavior of biting on hard objects, an equine vice related to boredom of confinement, is unknown in wild horses and occurs only in animals regularly tethered or stabled for considerable periods of time, the implication of such evidence is that horses were already at least tamed and controlled by Ice Age people (Leakey, 1981).

It is definitely known that relatively early in human history certain societies arose whose chief distinguishing characteristic was the centrality of the horse. One such nomadic equestrian people, the Mongols, conquered the most extensive empire in the history of mankind, largely through their skillful use of cavalry. Every Mongol warrior formed a close, intercommunicating unit with his superbly trained mount. The cavalryman could live on mare's milk supplemented by blood drawn from the vein of his horse. Marco Polo wrote that Mongol riders could remain mounted for days at a time, eating and sleeping on horseback (Gianoli, 1969). Although they became legendary as barbaric and merciless enemies in war, the Mongols are known to have treated their horses with care and devotion (Seth-Smith, 1978, p. 16).

Virtually all mounted societies from the beginning established a reputation for being fierce and imperious. On a practical level, of course, the animals gave riders overwhelming military advantage against pedestrian peoples. Mobility and power accruing from the mounted state were exploited to assure victory for those who possessed horses. The superiority bestowed upon a rider, however, was not only measured in palpable gain. For the psyche of equestrian people was deeply affected as well. Living intimately with the equine animal, caring for it, depending upon it for their way of life and their very existence, yet controlling it, exerting their will over it and incorporating its power as a physical extension of their bodies, had marked effect upon the mind. In mounted societies people not only merged physically with the horses that transported them, but intertwined their fate with the animals as well. Often they projected their sense of self into the human-horse relationship, establishing a strong feeling of identity with their mounts.

Unfortunately, an account of the effects upon humankind of riding the first horses, which, from archaeological evidence probably took place somewhere in the Ukrainian steppes, is lost to history. But horse peoples, from the Scythians and Assyrians of the ancient world, to the Cossacks of Russia, the Gauchos of Argentina, and the cowboys of the American frontier, whose cultures have been studied, are generally characterized as bold, fearless, aggressive, often proud and defiant. Using the North American Plains Indians as an example illustrates not only the swiftness with which the horse, once acquired by people previously unacquainted with the animal, can be completely incorporated into a society, but also shows the transforming effects of the animal upon the adapting culture. Several decades after becoming mounted, "lords of the Plains" became a suitable epithet for the nomadic tribes whose ferocity and daring as equestrians made them a scourge to their enemies. As early as 1803 a fur trader, Alexander Henry, noted the effect of the introduction of the horse upon the Indians, complaining that though horses "are useful animals," their acquisition had made the natives "indolent and insolent" (Coues, 1897).

Horses confer a feeling of heightened self-worth that is reflected in the behavior and perceptions of people who interact with them. And throughout much of human social history it has been the general rule that horses, both in a literal and symbolic sense, elevate the status of those who ride and use them. As Walter Prescott Webb wrote, "The horse has always exerted a peculiar emotional effect on both the rider and the observer: he has raised the rider above himself, has increased his power and sense of power, and has aroused a sense of inferiority and envy in the humble pedestrian. ... Through long ages the horse has been the symbol of superiority, of victory and triumph." The historian goes on to quote Lord Herbert: "A good rider on a good horse is as much above himself and others as the world can make him" (1936).

Its role in raising mankind to an exalted state, in addition to the many material benefits it alone provided prior to the machine age, earned considerable esteem for the horse in most societies. Equestrians are, literally and figuratively, "looked up to." Pride and arrogance go with being mounted. Contemporary usage perpetuates the linkage, for "get off your high horse" is a warning to the haughty, the meaning of which is clearly understood. From earliest times strong association has existed between horses and aristocracy, both in actuality and in symbol. If being astride a horse makes one feel like royalty, all the more reason that kings, princes, and lords must ride, while serfs and peasants walk. Riding horses has often symbolically separated the privileged from the lowly. Throughout history the ruling elite has frequently reserved horse-riding for itself alone. One thinks of Isak Dinesen's unforgettable story of the native African servant of a white settler, beaten so severely he has said to have willed himself to die, whose crime was riding on his master's horse when ordered to walk (1972).

A historical survey of horse-owning cultures of the world reveals a pervasive theme of high regard for the horse, which is almost universally recognized as the aristocrat among domestic animals, often identified with luxury, leisure, and power (Barclay, 1980, pp. 369–370). Evidence shows that beginning with the earliest civilizations human relationships with horses were especially meaningful. For example, elaborate burials in which as many as 29 horses were arranged in the grave around the body of their master give testament to the regard for the animals felt by the Scythians, who flourished in the fifth and fourth centuries B.C. (Trippett, 1974, p. 103). And an ancient Mesopotamian fable allows the horse itself to express the esteem in which it was held by society. Speaking to its friend, the ox, the horse boasts of "his pleasant life, how he is lodged near the King and great men, how choice and varied his food is and that his flesh is not eaten" (Aynard, 1972). Chivalry, embracing a complex of traits that represented ideal behavior for medieval noblemen, derives its name from the word meaning horse, and knights were paragons of equestrian skill. Historically, the cavalry has been the most aristocratic of the world's military units. The prestige that society grants to the horse persists into modern times, for the aura of equine grandeur remains to grace the machine age.

The fact that there are said to be more horses in America now, when their use is almost entirely restricted to pleasure, than there were during the age when the animals provided transportation expresses the appreciation that the present society has for horses. Interestingly, the old aristocratic association with horses is symbolically preserved in the formality and cultural status of many modern equine events. The fox hunt, for example, is a carefully perpetuated social ritual requiring strict adherence to prescribed behavior and clothing as well as mode of equitation. Horse shows are events often staged with the highest of elegance, where tuxedos and top hats set the tone for the exhibitors of magnificent saddle and fine harness horses. In racing, much personal status and prestige accrue to a victorious horse's owner, who shares in the animal's elevated rank as a thoroughbred that has proved its superiority by defeating its rivals in "the sport of kings."

From my recent field research with urban mounted police, it became clear that this unit of the police force is often viewed as "elitist," both by the officers themselves and by the general public. "A policeman on a horse is ten feet tall," officers say, and this brings special advantages. "Our unit is held in respect because the horse is a majestic animal." Often in the media, mounted policemen are given a heroic image as "Blue Knights." Mounted officers claim that foot and motorcycle policemen, who refer to them as "prima donnas and glory seekers," envy and resent the special status mounties have and the admiration they get from people on the street (Lawrence, 1985, pp. 116–173).

Ownership and use of pleasure horses in the United States today, of course, are not restricted to upper social and economic classes or to those who consciously strive for upward mobility. But people who ride and handle horses certainly seem to add to their self-image in a way that brings them a perceived elevation in status within their own particular social order. In addition to the satisfactions they bring to the present, horses symbolize and embody a dual heritage from the past. One influential component of American culture, especially in the East and South, is the old concept originating from British and northern European roots, which still identifies the horse with the leisured landed aristocracy (Barclay, 1980, p. 340). Undoubtedly of considerably greater importance in American consciousness is the more recent and pervasive egalitarian tradition of the Old West, represented by the cowboy as a New World horseman without noble lineage. For many people, the mounted cowboy herder continues to embody a longing for a life of freedom, equality of opportunity, autonomy, rugged individualism, adventure, challenge, and personal fulfillment in the less complex world of the bygone frontier (see Lawrence, 1984). As the cowboy's companion and essential partner, the horse becomes a means to recapture some of the cattle herder's glory. Rejecting the constraints of mechanized society to ride free over the plains in the wake of the cowboy hero becomes a way to recreate an experience of permanent cultural value. Although American pioneer philosophy left no room for rank emanating from aristocratic ancestry, it is ironic that cowboys soon set up their own standards to elevate themselves over and disparage those people who were excluded from their group. Actually classed as "hired hands," cowboys nevertheless developed an imperious attitude making them sometimes viewed as "the only reigning American royalty" (Cholis, 1977). It has been said that his horse made the cowboy a king, with the saddle as his throne. As one cowboy described his fellow cowpunchers, they "set themselves way up above other people who the chances are were no more common and uneducated than themselves" (Abbott and Smith, 1939).

Horses often serve as the fulfillment of nostalgic yearnings for the less complicated life of bygone eras. The motion and rhythm of the equine animal have always fascinated humankind and continue more than ever to do so in the mechanized age. Now that a practical need for horse transportation is secondary or nonexistent, people ride by choice, not necessity. A modern cowhand is still known to perform willingly any ranch task "if I can do it horseback," even when there are alternatives. A study of German immigrants to Brazil who adopted the horse complex soon after arrival in their new land revealed that they were very quickly transformed into people who "do not like to go afoot. Sometimes they lose half an hour or so with the rounding up of their horses in order to make a five minutes' trip to the neighbor's house" (Willems, 1944).

The number of people who ride for sheer pleasure is steadily increasing. Even those who do not ride, however, share appreciation of the special kinetic qualities of horses. People find equine motion a source of fascination and pleasure. Countless field-work informants describe the sense of comfort and well-being they experience upon hearing equine hoofbeats. City dwellers say they feel protected and safe to a much greater degree when they hear the clippety-clop of the mounted policemen's horses as compared to knowing there is a motorcycle officer or police cruiser nearby. People often express the idea that the rhythmic sounds remind them of the past, and many relate their sense of reassurance to the slower pace that is represented by the equine animal. The gait of the horse is a welcome natural rhythm in contrast to the tempo of the machinery in the modern motorized world and recalls an earlier time when human existence was more in harmony with the living environment (Lawrence, 1985).

Horses are intimately related to people's sense of the passage of time. Because of their rhythmic nature, equines have been closely associated with the marking out of time in human consciousness. One of the most persistent symbolic connotations of horses has been their role in drawing the chariot of the sun across the sky, enabling each day to come and go in predictable rhythm (Carr, 1965; Cooper, 1978; Howey, 1958; Neiman, 1979; Rowland, 1973). Ronald Blythe, whose portrait of rural life in Akenfield often centers on the changes in people's lives that came with mechanization, articulated the horse's role in demarcating time. "Nothing has contributed more to the swift destruction of the old pattern of life in Suffolk than the death of the horse. It carried away with it a quite different perception of time" (1969). J. Frank Dobie observed, "the more machinery man gets, the more machined he is. When the traveler got off the horse and into a machine, the tempo of his mind as well as of his locomotion was changed" (1952). E. T. Hall described his own sense of time being affected by a horseback journey in which he was "in the grip of nature." After a few days of adjusting to the rhythm of the horse's gait, he reveals, "I became part of the country again and my whole psyche changed. ... The urban tempo," he points out, is "out of sync with the body" (1983).

Society makes use of the symbolic association of horses with the passage of time. For example, a 1984 television advertisement for Budweiser beer shows a horse dramatically in motion across the open landscape while a voice explains the great amount of time the company takes in order to brew beer of the finest quality. The advertising sequence, still centered on the rhythmic movements of the horse, ends with the assertion that because of the slow aging process used in its brewing, Budweiser is the best beer to drink—"time, after time, after time."

The marking out of time is "one of the fundamental applications of order, for no communal human activity can take place without it" (Canetti, 1978). The regulation of time binds a group of people together as a society, making possible their participation in shared activities. Thus, horses, given their association with time, can represent the social order itself. Horses participate in parades, which generally celebrate the seasons, the passage of certain segments of time, or commemorate historical events. When the use of old tradition is desired for important events in society, a slow pace and measured tread is imposed to denote solemnity and formality. Thus, horses are often designated as mounts for honor guards or escorts for visiting dignitaries, and their presence at special civic and national occasions like military funerals lends an air of formality not provided by machinery. In such modern ceremonial and state functions horses represent the historic past and lend the sanction of culture and tradition in symbolizing the social order (Lawrence, 1985, pp. 148–149). Horses are often said to stand for war and conquest. This societal connotation is easy to understand, since the animals were chief instruments of battle throughout such a large segment of recorded history, and most cavalries were officially abolished only as recently as the 1940s. That a large, gentle herbivore, whose natural reaction to danger is swift flight, can be made to gallop forward into the noise and confusion of battlefields is testament to equine obedience to the human will. Herd instinct and adherence to equine social dominance orders may play a part in the cavalry charge (Barclay, 1980, p. 358; Clutton-Brock, 1981). But if this is the case, the rider in a sense takes the place of the horse's conspecifics, and conditioning and human mastery are superimposed upon the animals (see Lawrence, 1985, p. 158).

The meaning of the horse as a metaphor for man's conquering force rests on the knowledge that the horse upon which the victorious warlord rides has itself been previously conquered. For each individual mount has first to be "broken" and trained to make it suitable for human use. The wild-to-tame transformation must be reenacted in the process of creating every equine working partner. The inherent wildness of horses is a symbolic concept often referred to in daily life, and the equine species is known to have been more resistant to taming than many other domesticated animals (Zeuner, 1963, p. 329). "It requires no mean skill on the part of man to assert his dominance over such an animal [wild horse] and break it in for riding" (Clutton-Brock, 1981, p. 86). Domination over the animal metaphorically sets the stage for further conquest. Bucephalus, the spirited war-horse ridden by Alexander the Great, had defied the mastery of all the experts in his father's kingdom of Macedonia before the youth tamed him. It was Alexander alone who could conquer the horse through overcoming the animal's fears, thus making him a fitting partner for some of the greatest feats of worldly conquest known to history (Plutarch, 1980).

Human interaction with horses, of course, has its dark side. Instances abound in which the debt society owes to the horse has not been fairly paid. Cavalry mounts who once served their country valiantly have been abandoned to exploitation involving extreme brutality in foreign lands (Cooper, 1983). Owen Wister's first-hand account of a horseback trek in the frontier West revealed an example of unspeakable atrocity perpetrated by a rider who, unable to satisfy his anger by merely beating his horse, gouged out the eye of the disobedient animal (1958). Cart horses, drawing human burdens throughout the ages, have been frequent victims of savage cruelty. Though coach horses are now obsolete, their abused and malnourished modern counterparts in some cities still draw vehicles to artificially recreate the past for the amusement of the tourists. And society must bear the guilt for such current realities as the drugging of race horses and the torture inflicted to produce artificial gaits in certain breeds.

Horses, with their sensitivity, innocence, and unusual willingness to give themselves fully to the tasks society demands of them, seem particularly vulnerable to human neglect and cruelty. History is replete with examples of cultures that have practiced animal sacrifice to appease their gods, and the high value assigned to horses has made them prime subjects for ritual bloodshed. In *Equus*, the main character perceives horses as interchangeable with the crucified Christ, and blinding the animals, as he does in the play, becomes the ultimate act of violence toward the equine sacrificial substitutes (Shaffer, 1974). Barbarity toward the powerful yet submissive equine animal is often peculiarly expressive of human malice. De Maupassant's story of Coco, the old and faithful work horse, who is systematically beaten and starved to death by a vengeful youth, is a classic representation of sadism (1981). Often, the theme of the suffering of horses is metaphoric for the trials of human beings who become identified with the animals as victims of the same inequitable social system. Tolstoy immortalized the commonality of pain and travail between horses and people in his story of the horse, Strider (1972). When this tale was produced as a play in 1980, a man wearing a horse-collar and bridle took the part of Strider, and a musical solo sung during the drama reiterated the theme "Oh hard is life for man and horse."

The hard labor of horses was the power that gave rise to nations. When the Spanish conquistadors met with success against their pedestrian foe in their military expedition to the New World, they acknowledged that "after God, we owed victory to our horses" (Graham, 1930). Most Americans agree that the horse enabled the United States to grow into the great nation that it became. Four distinctly American breeds were recently honored on commemorative postage stamps, because horses were "so instrumental in the development, exploration, and expansion of our nation" (McClung, 1986). Fulfillment of Manifest Destiny required that people be mounted, for the horse was the essential instrument by which conquest of the wilderness and settlement were made possible. As Owen Wister has asserted, it was not just the American pioneers' special traits, but also the destiny that brought the Anglo-Saxon colonists into partnership with a particular kind of horse, the mustang, as their "foster-brother" and "ally," which determined the course of history on the American continent (1972).

Mustangs are considered to be descendants of the domesticated mounts that were brought to the New World by the Spanish and then reverted to the feral state. Today, some mustang herds still run free in certain areas of Western rangeland. But their continued existence is called into question because of sharp dis-agreement among opposing segments of contemporary society. Bitter controversy divides opinions about the feral horses' value and significance. Those who want them removed from the range say that the mustangs serve no useful or practical purpose, yet they compete with cattle and domestic horses for grass and water and use land that should be reserved for wildlife, especially the game species exploited by hunters. Since the horses are feral and not native wild animals, their detractors claim that they upset the natural balance of the ecosys-tem.

Defenders look upon the feral horse as living symbols of the historic and pioneer spirit of the West, an aes-thetically vital and cherished part of the nation's heri-tage. They point out that even though native wild horses became extinct in North America as a result of some undetermined cause operating just after the last Ice Age, they had previously evolved as a species on this continent and therefore are still ecologically com-patible, posing no real threat to indigenous wildlife. Thus, the equine animal is the focus for conflicting American societal value-systems applied to man and nature—the economic versus the aesthetic, the prag-matic versus the affective, the tame versus the wild, and the anthropocentric ethos versus belief in the intrinsic right of all forms of life to survival.

Horses are powerful worldwide symbols, and when-ever the animals have become significant in human society, people have typically taken inordinate pride in projecting the proper dignified image of themselves as equestrians. Simple transportation has seldom been the only issue involved. If pure expediency were the deter-mining factor, for example, certain African populations would have ridden quaggas instead of horses. Until the 1840s, the quagga, a species of equid, was common in south and central Africa. It was docile, easily tamed and trained for riding, sturdy, and well-adapted to its environment. Moreover, quaggas were plentiful and readily available, in contrast to costly foreign horses, which had to be imported by sea. Instead of being used for riding, however, quaggas were hunted to extinction. English and Boer farmers viewed quaggas more as vermin than as potential mounts, and their self-image did not allow for the riding of an animal that was classified as wild and whose coarse appearance offended their aesthetic taste. Motivated largely by a deeply inculcated belief that only true horses represented status, colonists resisted the obvious utilitarian advantages of using quaggas as riding mounts and chose instead to import horses (Downs, 1960).

The term most universally and frequently applied to horses is "noble." Again and again people reiterate the concept of the horse as the noblest of domesticated animals. Somehow the horse's perceived traits—dignity and refinement, grace, beauty, and power—are felt to be transmitted to its human associates, for the equine animal seems to confer self-esteem and even ennoblement. People of horse-owning societies are prone to take not only their own image, but also their ethnic identity, from the horses that so typically become the focus of their cultural life.

Horses, once acquired by the North American Plains Indians in the eighteenth century, for example, quickly transformed their way of life and soon became the standard measurement of all that was held of value by society. A tribesman with few horses was a pauper who "trudged afoot" (Lowie, 1963). Today, contemporary Plains natives like the Crows continue to identify with this heritage and say that to be on horseback makes them feel truly Indian. They still find in their horses a distillation of the qualities that are most meaningful in native culture. The animal is important in helping them define themselves as a social entity and in retaining an important part of their heritage despite encroachment by the dominant society (see Lawrence, 1985, pp. 24–54).

Contemporary Crow Indians often compare themselves to Arabs, who traditionally consider their horses as family, and empathize with the idea that "love for horses flows in Arabic blood." It is legendary that in Arabia "horses are riches, joys, life, and religion." Arabs have a sacred duty to give horses good care and are charged for the love of God not to be negligent toward them, lest they "regret it in this life and the next." According to the Prophet, God created the peerless Arabian horse from the wind, and told it "I have hung happiness from the forelock which hangs between your eyes; you shall be the lord of the other animals. Men shall follow you wherever you go. ... You shall fly without wings; riches shall be on your back and fortune shall come through your mediation." (Daumas, 1968).

In other societies, too, dependency upon the horse led to an attitude of appreciation in which identity became invested in the animal. Traditional Gypsies, for example, feel that the nomadic existence made possible by the horses who draw their caravans is the only way of living worthy of man (Clebert, 1963, p. xvii). For them, "life revolves around the horse," and greeting between friends is not "I hope you will live happily" but rather "May your horses live long" (Seth-Smith, 1973, p. 312; Clebert, 1963, p. 102). The truism that "a Gypsy without a horse is no genuine Gypsy" (McDowell, 1970; Clebert, 1963, p. 103; Erdoes, 1959) indicates the extent to which the animal has gone beyond utility to become essential to group identity. Horses represent not only the sense of self but also embody intense social meaning. The role of the horse is central to the Gypsy lifestyle, enabling people to move from place to place as their culture dictates. Traditional nomadism sets these people apart both physically and ideologically from the constraints of sedentary living, which they abhor. Their strong views of settled peasants and farmers as beneath them are comparable to those of American range cowboys who also hold in disdain agriculturalists and other people whose labors make them pedestrians, asserting that "a man afoot is no man at all." Self-identity and group belonging both depend upon the horse.

Today, though mechanization has made the historic utility of horses for traction and transportation virtually obsolete, and societies no longer revolve around them, many contemporary people still turn to horses for important functions. Horses are companions of a different order than are most other domestic species, for they provide an experience of motion that defies machinery. Equestrians can merge their own being with the rhythm and power of their mounts, bonding with nature in a participatory rather than a passive way. Fine-tuned intercommunication between rider and horse is both physical and mental. The beauty and grace of movement of the horse becomes something possessed by the rider.

There is innocence in the horse's nature, enhanced by the strange paradox of its great strength combined with tractability. The horseman's expression that a certain animal has a "kind eye" encapsulates the feeling for a creature of enormous power in whom such benevolence toward people can still be found. Horse sense is a highly regarded quality, and stories of equine sagacity commonly feature faithful and heroic horses who serve and rescue their masters. Jonathan Swift in *Gulliver's Travels* described horses in his native England as "the most generous and comely animals we had" (1985). It was the equine species that he chose in depicting the harmonious land of the Houyhnhnms, paragons of rationality, wisdom, and nobility, in order to satirize the society of his day, made up of Yahoos, human beings whose illogical behavior, vices, follies, and degradations contrasted so sharply with the virtues of the horses. The immutability of equine characteristics—obedience, mildness, and non-aggression—is used metaphorically to represent the set and natural order of the world. In *Macbeth*, for example, Shakespeare uses the image of tame horses, sedate and disciplined, suddenly going wild to the extent that the once-gentle herbivores devour each other's flesh (1972). This frightful equine transformation dramatizes the enormity of the crime of murder, which has not only broken the human social code, but has disturbed the very order of nature itself.

On a symbolic as well as a physical level, horses transport people to new places in body and mind, carrying them out of the mundane world into the richer realm of imagination. Beginning in childhood, they provide the joy of motion as rocking horses and carousel steeds, which can be ridden to the fulfillment of splendid dreams. A hobby-horse gives its name to an absorbing interest that metaphorically carries a person away from a humdrum existence, so that even the self is left behind. Pegasus, the horse with wings, is symbolic of poetic inspiration that surpasses the ordinary and transcends earthly roots. In certain societies shamans, as healers, use a symbolic horse that carries them on a mystical journey away from the everyday world, beyond time and space, into the sacred domain. By means of figuratively riding a horse, it is possible to reach heaven, a sphere inaccessible to ordinary people, where they can be empowered for curing (Eliade, 1974).

Perhaps the best way to describe the horse's role in modern society is as a healer. As a companion who shares our leisure, it provides exercise, diversion, beauty and grace, and a sense of communion with nature, which may return us to our roots and restore us to a sense of harmony. It is often patient and kind, with a judicious sense of the character of its rider, for a horse is willing to modify its behavior for the young or the handicapped. An old adage states that "the best thing for the inside of a person is the outside of a horse." Folklore, in which it seems even many modern physicians believe, though they cannot explain it, holds that association with horses leads to longevity. Mechanization still has not removed from the human consciousness the concept of measuring physical force by "horsepower." Somehow, the utility of machines has never really supplanted horses, perhaps because what was said in the fourteenth century by the Duke of York still rings true for modern men and women:

Men are better when riding, more just and more understanding, and more alert and more at ease, and more under-taking, and better knowing of all countries and all passages; in short and long all good customs and manners cometh thereof, and the health of man and of his soul" (Vesey-Fitzgerald, 1965).

References

Abbott, E. C. (Teddy Blue), and H. H. Smith. 1939. *We Pointed Them North: Recollections of a Cowpuncher*, p. 247. New York: Farrar & Rinehart.

Aynard, J. M. 1972. Animals in Mesopotamia. In *Animals in Archaeology*, ed. A. Brodrick, p. 65. London: Barrie & Jenkins.

Barclay, H. B. 1980. *The Role of the Horse in Man's Culture.* London: J. A. Allen.

Blythe, R. 1969. *Akenfield*, p. 18. New York: Pantheon.

Canetti, E. 1978. *Crowds and Power*, p. 397. New York: Continuum.

Carr, W. G. 1965. *Man and Animal: Man Through His Art*, vol. 3, p. 48. Greenwich, CT: New York Graphic Society.

Cholis, J. 1977. John Wayne, Cattleman. *Persimmon Hill* 7: 28–35.

Clebert, J. P. 1963. *The Gypsies*. London: Vista.

Clutton-Brock, J. 1981. *Domesticated Animals from Early Times.* Austin: Univ. of Texas Press.

Cooper, J. C. 1978. *An Illustrated Encyclopaedia of Traditional Symbols*, p. 85. London: Thames and Hudson.

Cooper, J. 1983. *Animals in War*, p. 47. London: William Heinemann.

Coues, E., ed. 1897. *New Light on the Early History of the*

Greater Northwest: The Manuscript Journals of Alexander Henry and David Thompson, 1799–1814, 3 vols., vol. I, p. 225. New York: Francis P. Harper.

Daumas, E. 1968. *The Horses of the Sahara*, pp. 7, 28, 29, 31–3. Austin: Univ. of Texas Press.

de Maupassant, G. 1981. Coco. In *The Book of Horses*, ed. Fred Urquhart, pp. 15–9. New York: William Morrow and Co.

Dinesen, I. 1972. *Out of Africa*, pp. 278–83. New York: Vintage Books.

Dobie, J. 1952. *The Mustangs*, p. xiii. Boston: Little, Brown.

Downs, J. F. 1960. Domestication: An Examination of the Changing Social Relationships Between Man and Animals. *Kroeber Anthropological Society Papers* 22: 18–67.

Downs, J. F. 1961. The Origin and Spread of Riding in the Near East and Central Asia. *American Anthropologist* 63: 1193–203.

Eliade, M. 1974. *Shamanism*, pp. 173–74, 467. Princeton: Princeton Univ. Press.

Erdoes, K. 1959. Gypsy Horse Dealers in Hungary. *Journal of the Gypsy Lore Society* 38(1–2): 1–6.

Gianoli, L. 1969. *Horses and Horsemanship Through the Ages*, p. 71. New York: Crown.

Graham, R. B. 1949. *The Horses of the Conquest*, p. 11. Norman: Univ. of Oklahoma Press.

Hall, E. 1983. *The Dance of Life*, pp. 39–40. Garden City, NY: Doubleday.

Harris, M. 1985. *Good to Eat: Riddles of Food and Culture*, pp. 96–7. New York: Simon and Schuster.

Hope, C. E. G., and G. N. Jackson, eds. 1973. *The Encyclopedia of the Horse*, p. 236. London: Peerage Books.

Howey, M. O. 1958. *The Horse in Magic and Myth*, pp. 114–25. New York: Castle Books.

Lawrence, E. A. 1984. *Rodeo: An Anthropologist Looks at the Wild and the Tame*, pp. 49–82. Chicago: Univ. of Chicago Press.

Lawrence, E. A. 1985. *Hoofbeats: Studies of Human-Horse Interactions*. Bloomington: Indiana Univ. Press.

Leakey, R. E. 1981. *The Making of Mankind*, pp. 193–96. New York: E. P. Dutton.

Lowie, R. H. 1963. *Indians of the Plains*, p. 44. Garden City, NY: Natural History Press.

McClung, G. S., ed. 1986. Horse Sense (woman whose idea stimulated the horse commemorative postage stamps). *Mount Holyoke Alumnae Quarterly*, vol. LXX, no. 1: 43.

McDowell, B. 1970. *Gypsies: Wanderers of the World*, p. 103. Washington: National Geographic Society.

Neiman, L. 1979. *Horses*, p. 321. New York: Abrams.

Plutarch. 1980. *The Age of Alexander*, pp. 257–58. New York: Penguin Books.

Rowland, B. 1973. *Animals With Human Faces: A Guide to Animal Symbolism*, p. 110. Knoxville: Univ. of Tennessee Press.

Seth-Smith, M., ed. 1978. *The Horse in Art and History*. New York: Mayflower Books.

Shaffer, P. 1974. *Equus*. New York: Atheneum.

Shakespeare, W. 1972. *Macbeth*, p. 73–5. New York: Amsco.

Swift, J. 1985. *Gulliver's Travels*, p. 287. New York: Penguin Books.

Tolstoy, L. 1978. Strider. In *The Portable Tolstoy*, ed. John Bayley, pp. 435–74. New York: Penguin Books.

Trippett, F. 1974. *The First Horsemen*. New York: Time-Life Books.

Vesey-Fitzgerald, B., ed. 1947. *The Book of the Horse*, p. 26. Los Angeles: Borden.

Vesey-Fitzgerald, B., ed. 1965. *Animal Anthology*, p. 30. London: Newnes.

Webb, W. P. 1936. *The Great Plains*, p. 493. Boston: Houghton Mifflin.

Willems, E. 1944. Acculturation and the Horse Complex Among German-Brazilians. *American Anthropologist 46*: 153-61.

Wister, F. K., ed. 1958. *Owen Wister Out West: His Journals and Letters*, pp. 108–09. Chicago: Univ. of Chicago Press.

Wister, O. 1972. The Evolution of the Cow Puncher. In *My Dear Wister: The Frederic Remington-Owen Wister Letters*, ed. Ben M. Vorpahl, p. 81. Palo Alto: American West.

Zeuner, F. E. 1963. *A History of Domesticated Animals*. New York: Harper & Row.

The Animal as Alter Ego: Cruelty, Altruism, and the Work of Art

Marianna R. Burt

In literature, the alter ego is a well-known device, a similar but separate character (literally, a "second self") through which the central figure may play out or reinforce the implications of his or her acts. Freud, citing Macbeth and Lady Macbeth, showed how two dramatized beings could constitute a single personality; students of the human-animal bond might wonder if this is also true of people and their animal companions. Burt explores the question through examples ranging from Greek vase painting to the modern novel. The Andokides amphora (c. 510 B.C.), depicting the meeting of Heracles and Cerberus, sets the terms of the discussion, followed by examinations of many later portraits in which a companion animal resembles the human sitter in feature, pose, or expression. Artists such as Hogarth, Gainsborough, Buhot, and Eakins often depicted pets as doubles of their owners, thereby illuminating the qualities of each.

In literature, the extended narrative structure permits even more complexity. Events as well as characters can be doubled, and writers as diverse as Dostoevsky, Camus, Faulkner, and Richard Wright have used animal doubles to shape the flow of their work. Cruelty and altruism toward animals take on new meaning in such contexts. The doubled characters are often in conflict, and the animals they befriend or abuse serve as the foci of their own self-love or -loathing. Some works describe a resolution of the conflict, others a continued fragmentation of personality.

Burt concludes her analysis with a reexamination of the Andokides amphora as a symbol of Jung's dictum that man, to understand himself, must befriend the animal within himself. EDITOR

The alter ego, or double, is a potent theme in many works of literature and art. The idea of a separate self taps the wellsprings of the creative imagination, and modern works that employ this device relate back to the ancient mysteries and magic of myth.

There are striking double portraits, such as Mary Cassatt's *The Loge* (1882), that push affinities to the point of replication. Van Gogh found a double in the physician Paul Gachet, who supervised him at Auvers during the last few months of his life. In June 1890, Van Gogh painted two portraits of Gachet that resembled closely a major self-portrait he was then completing. He wrote to Gauguin, whom he also viewed as an alter ego, "I have painted Dr. Gachet with the heartbroken expression of our times" (Rewald, 1962).

There is no dearth of examples in literature. Freud pointed out that Macbeth's feelings are expressed and elaborated in the behavior of Lady Macbeth (Freud, 1964). The most crystalline example may be the doubling of Hector and Achilles in the *Iliad*, which goes far beyond similarities in ability and rank. Patroclus, it will be remembered, donned the armor of Achilles to trick the Trojans, only to be killed by Hector. The victor stripped the armor from Patroclus's body and wore it in combat with the avenging Achilles. Achilles, filled with guilt for permitting the deception that killed his friend, had to face and destroy the Trojan hero costumed as himself—his very image, his double.

Given the abundance of works of art depicting human-animal relationships, it is not surprising to find a substantial number in which affinity approaches identity and the animal doubles for the human character. In Figure 1, the amphora by the Andokides painter

*Figure 1. Andokides Painter, Heracles and Cerberus. Amphora,
c. 510 B.C., Musée du Louvre, Paris.*

(c. 510 B.C.) shows such an encounter between Heracles
and the many-headed watchdog Cerberus, the guardian
of the gates of Hades, whom he must carry up to earth
to complete his final labor. While all his other labors
have been accomplished through brute strength, Heracles is expressly forbidden to use weapons to capture
Cerberus. The ancient vase painter depicts a scene in
which the hero befriends the animal, extending a hand
to touch an unlifted head. The arc formed by their bodies is echoed by the curved lines in the setting, which
join the figures in a dyad, creating psychological distance from the goddess Athena on the left.

   There are many parallels between the driven Greek
hero and the fierce watchdog of the dead, but the
strongest visual expression of their doubling lies in the
fact that Heracles, too, appears to have more than one
head. Across his shoulders, he wears the famous mantle
made from a leopard he has slain. This unique align-
ment, and its symmetry, forges a significant bond between the figures.

For myth, the consequences of human-animal doubling are very great. One effect is the use of animal masks and impersonation in religious ritual, theatre, and dance. Linked with this is the elaboration of fantastic human-animal hybrids, which were often worshipped, as well as the growth of bestiality myths, such as that of Leda and the swan.

Yet animal doubles may be based upon a purely naturalistic stressing of resemblances. Many important portraits confirm the popular wisdom that people and their pets often look alike, in much the same way that husbands and wives are said to grow to resemble each other with the passage of time. The artist will highlight the special qualities of the bond through color and composition. In Renoir's *Madame Charpentier and her Children* (1878), it is no accident that the sitter wears a black dress with white border, the same color pattern seen in the fur of the family dog. These two figures stabilize the composition; the dog's horizontal placement balances the strong downward thrust of other elements in the room. Indeed, it might be said that the pet holds the family together—he prevents the children from tumbling out of the picture plane!

In *Miss Jane Bowles* (1775), Sir Joshua Reynolds so entwines the shapes of the little girl and her dog that it is hard to tell where one ends and the other begins. By placing the dog's head directly below the child's, he makes it easy to discern an actual similarity of facial shape and feature. In Gainsborough's *Mrs. Perdita Robinson* (1785), the dog, who shares his mistress's coloring, stands devotedly at her side, the line of his chest actually blending into the curve of her arm. In *Sleeping Girl*, also known as *Girl with a Cat* (1880), Renoir's young subjects have the same small mouth and pert, upturned nose. Color orchestration of fur and costume emphasizes their doubling, continued to the point of placing the cat's paw within the girl's identically shaped hand.

Yet the most explicit human-animal double portrait must surely be Hogarth's *Self-Portrait* of 1745 (Figure 2), in which the artist yields the foreground to his beloved pug, Trump. Hogarth's contemporary, Samuel Ireland, recalled, ". . . It had been jocularly observed by him that there was a close resemblance between his own countenance and that of this favorite dog, who was his faithful friend, and companion for many years, and for whom he had conceived a greater share of attachment than is usually bestowed on these domestic animals . . ." (Burt, 1985). The physical resemblance between the pug's face and that of Hogarth, with his square jaw and upturned nose, is obvious, but the doubling extends far beyond that. While, at first, the oval frame behind the artist, a standard device in portraiture, appears to set him apart from the rest of the canvas, in fact, the opposite is true. The oval also suggests a mirror, and the way in which dog and man are turned toward each other, as well as the compression of space between them, further evokes a mirror image. The same pattern of brush strokes is used for Hogarth's drapery and the dog's chest area; except for color difference, one seems a continuation of the other. The curves of Trump's body are repeated in the artist's shoulder. At a time when portrait sitters were often depicted with their most valued possessions, Hogarth chose to dramatize his bond with his beloved pet through striking affinities of form.

In art, many cases of human-animal doubling occur in what we may call situational doubles, rather than portraits. Subjects are not so much individuals as types, and their roles are largely narrative. Often, they are found in genre paintings such as Pieter de Hooch's *The Fireside* (c. 1670–75), where the little girl at the lower right holds a cat whose round, over-large head is like her own (Figure 3). As if jokingly to confirm this doubling, the child holds the cat's paw in her hand—and the angles of its foreleg and her lower arm are the same. At times, the narrative role can be quite explicit. In

Figure 2. William Hogarth, Self-Portrait *(detail). Oil on canvas, 1745, The Tate Gallery, London.*

Figure 3. Pieter de Hooch, The Fireside. *Oil on Canvas, c. 1670–75,*
North Carolina Museum of Art, Raleigh.

Sympathy (1877), Breton de la Rivière's child and dog
share outlines and repeat each other's contours, suggesting interdependence and comfort in an unhappy
moment (Figure 4).

Situational doubling plays a major role in the work of
the great printmaker Félix Buhot. With Buhot, animal
figures often serve as accents, repeating the actions and
reinforcing the feelings of the human subjects. Consider
two very different rainy scenes: In *The Celebration*, a
jaunty, bouncing dog restates the rollicking movement
of the couple he accompanies, while, in *The Night
Walker*, a desolate human form is surrounded by a
corresponding cluster of society's refuse—a group of
stray animals.

Figure 4. Breton de la Rivière, Sympathy. *Oil on canvas, 1871, in the collection at Royal Holloway and Bedford New College, University of London, Egham, Surrey.*

Figure 5. Félix Buhot, Winter in Paris, *1879. Etching, aquatint, and dry-point, fifth state, by the courtesy of The Boston Public Library Print Department.*

In Buhot's masterpiece, *Winter in Paris, 1879* (Figure 5), human and animal figures together capture the atmosphere of that brutal season and stress the suffering that was then commonplace. The fallen horse in two of the remarques is but the culminating image of these hardships.

The centerpiece is a complex Paris street scene glimpsed through several doubled relationships. To the left, the elegantly dressed society woman and her poodle maintain a distance, hurrying away from the scene of hungry, foraging dogs and shabby, shivering men at the lower right. Only the curiosity of the turning child links them with the rest of the scene. The little girl, poodle, horses, and left background figures form the suggestion of a circle around an enigmatic, solid black figure at the center of the scene, where the vanishing point occurs. This is the artist as observer—and we find his double in the small black dog looking up at him from our right. It is this dog that has attracted the child's attention, hinting that all areas of life are accessible to the artist's inquiry.

In literature, too, there are many situational human-animal doubles. In Steinbeck's *Of Mice and Men* (1937), old Candy's ancient dog is his double, and his helplessness when the other men decide that the dog must be killed foreshadows George's unwilling mercy killing of Lennie at the close of the tale. When the dim-witted Lennie accidently crushes a puppy, this sets the stage for his unintended killing of another creature he sees as soft and cuddly—Curley's wife. In fact, in death, the woman and the puppy lie side by side in the hay.

In Dostoevskyy's short story, "The Double" (1972), the misanthropic central character encounters and rebuffs a stray dog—a symbol of his own present and future displacement from society—just moments before he first spies his human double, the man who has stolen his identity. But this is only a passing allusion, to be developed more fully in Dostoevsky's other writings. In contrast, the animal double is a true pivotal factor in such works as Steinbeck's *The Red Pony* (1959) and Rawlings's *The Yearling* (1938).

In both of these works, the doubling of child and animal is fully articulated. The lonely boy identifies with the creature, which becomes his special responsibility. The relationship, even though doomed, is an essential part of the passage to manhood; in each tale, the boy cares deeply about a creature other than himself and must survive its loss. We read at the end of *The Yearling*:

> *Flag—He did not believe that he should ever again love any-thing—man or woman or his own child, as much as he loved the yearling. He would be lonely all his life. But a man took it as his share and went on.*
> *In the beginning of his sleep, he cried out, "Flag!"*
> *It was not his own voice that called. It was a boy's voice.*

Somewhere beyond the sink-hole, past the magnolia, under the live oaks, a boy and a yearling ran side by side, and were gone forever.

In both novels, the death of a young, trusting animal brings about another death—that of boyhood and innocence. Many other twentieth-century books for younger readers employ variations of this theme, often with a happy ending in which resolution of the animal's plight is paralleled by integration of the lonely adolescent into the social environment (Burt and Harding, 1986).

But there is a darker side to human-animal doubling, as we see in Camus's *The Stranger* (1957). Meursault's wretched neighbor, Salamano, has an equally miserable dog as his double:

> *The spaniel had a skin disease, mange I think, which made him lose almost all his hair and covered him with sores and brown scabs. On account of living with him, the two alone in a small room, old Salamano ended up looking like him. There were red scabs on his face, and he had thin, yellow skin. The dog, for his part, had adopted from his master a kind of bent way of walking, his muzzle forward and his neck stretched out. They had the look of being related, yet they detested each other.*

Camus evokes the absurd image of Salamano's dog in Meursault's final outburst on "the benign indifference of the universe." Its irony emerges from the man's routine mistreatment of his dog, juxtaposed with his acute misery when it disappears. Such human-animal doubling in a deeply conflicted personality allows self-hatred to be vented on an animal with which the tormented person feels a particular affinity. This is pushed to its logical conclusion in Thomas Mann's short story, *Tobias Minderkindel* (1936), in which the meek, cowardly Tobias, who is often bullied, alternately loves and punishes a timid dog he has rescued from cruelty, until, at last, he kills it.

"Killing is but one aspect of the wandering grief we endure," wrote Rilke in Canto XI of *Sonnets to Orpheus* (1961), referring in part to the practice of catching wild birds in the Kurst caverns near Duino, Trieste. Perhaps the most powerful depiction of the paradox of cruelty and self-punishment is Raskolnikov's dream and its aftermath in Dostoevsky's *Crime and Punishment* (1975). The student Raskolnikov, in the grip of a compulsion to rob and murder an old pawnbroker, has a detailed, horrifying nightmare in which he watches helplessly while the villagers of his hometown beat an old carthorse to death. Their mounting mood of violence, which sweeps away all reason, is, of course, akin to Raskolnikov's inability to fight off his compulsion. Just as the mare loses her strength under the torrent of blows and sinks to the ground, Raskolnikov's strength deserts him and he wakens with the feeling that he is doomed to carry out the crime: Raskolnikov himself is condemned, not to die, but to act as executioner. In this convolution of violence, the decrepit animal, with which he had sided in the dream, is replaced by the old woman and her sister, who become his victims. The murder scene takes place in the same helpless, dreamlike atmosphere that pervades the nightmare. Raskolnikov behaves like an automaton (or a sleepwalker), even during his most violent actions. Animal metaphors used in the murder scene sustain the doubling effect of the dream. The mean, ugly old woman seems scarcely human; her sister Lizaveta, who returns unexpectedly to the flat, is "simple, crushed, and cowed," mute as a sacrificial animal, not even lifting her hand to ward off the blow that kills her.

Going even further, we note that the same Russian words and phrase patterns are used to describe key moments in the killing of the old horse and the murder of the two sisters. Both Lizaveta and the horse are called *wretched*. The mare is dealt a *crushing blow*, (literally, *one that crashes down*), and she sinks to the ground; when the *blow* falls on the woman's skull, she *crashes down* instantly. The peasant Mikolka swings the shaft with both hands over the mare and brings it down upon her *with all his might*; these very words are used to describe Raskolnikov's blows to the skull of the pawnbroker. Interestingly, as if to restate the paradox of Raskolnikov's dual role as executioner and victim, the animal's last struggle, in which she *staggered up, tugging, tugging with her last strength*, becomes for the murderer an ordeal in which *his strength seemed to have deserted him*. When he tries to remove the cord from the pawnbroker's neck, he *tugged at it, but it was too strong to snap*.

The parallel is complete when Raskolnikov labors like a beast of burden in the Siberian prison camp. He, too, experiences death—a serious illness during which he has hallucinations and terrifying dreams. This is followed by rebirth in the form of a religious conversion.

For Dostoevsky, the conflicted personality is at war, not only with himself, but with the rest of the universe, and this frequently is manifested in his relations with animals. In *The Brothers Karamazov* (1950), the village idiot who crushes the goose's neck beneath the cart foreshadows another "idiot," the epileptic Smerdyakov, who murders Fyodor Karamazov.

In this novel, an important instance of human-animal doubling occurs in the subplot of the schoolboy Ilyusha. This boy, who is often bullied by his schoolmates, becomes cruel in turn and feeds a stray dog a morsel containing a pin. When he falls fatally ill, he believes that he is being punished for his misdeed, a conviction shared by several other characters regarding the consequences of their own forms of cruelty. The boy's happy discovery that the dog he abused is still alive and has been adopted—that is, has a place in the social order—echoes Dmitri Karamazov's realization that he is not responsible for his father's murder.

For Dostoevsky, the link between animals and man goes beyond doubling to form part of an all-encompass-ing system of correspondences constituting the order of the universe. In *The Brothers Karamazov*, Father Zossima tells his followers that "all is like an ocean; all is flowing and blending; a touch in one place sets up movement at the other end of the earth." Rather than being lower than man, animals are, in a sense, above him, since man alone is sinful:

> "Look," said I, "at the horse, that great beast that is so near to man; or at the lowly, pensive ox, which feeds him and works for him; look at their faces, what meekness, what devotion to man, who often beats them mercilessly. ... It's touching to know that there's no sin in them, for all, all except man is sinless, and Christ has been with them before us."

Under such a worldview, an altruistic act, whether toward animal or human, is, above all, psychologically sound, being an expression of the "all-embracing love" that binds us to one another: ". . . in truth we are each responsible to all for all, it's only that men don't know this. If they knew it, the world would be a paradise at once." All forms of cruelty are thus profound violations of the moral order, and Father Zossima admonishes his readers: "Love the animals . . . don't harass them, don't work against God's intent." Significantly, the novel ends with the funeral of the boy Ilyusha; once cruel and alienated from his fellows, he is loved by all before he dies. In dying, he asks that bread be sprinkled on his grave so that sparrows may feed there; for their part, his comrades vow that they will "always remember the poor boy at whom we once threw stones." A healing reconciliation has been achieved, emblematic of the resolution of other conflicts within the novel.

No such affirmative vision exists in the world of black writer Richard Wright, who often uses animal doubles to heighten his stories' pervasive sense of despair. His characters behave violently toward animals largely because of the prolonged brutalization that they and their social environment have undergone at the hands of whites. In *Black Boy* (1945), the central figure is disgusted with himself when he tortures a crayfish, and "awareness of his cruelty in this incident announces a life of self-loathing to come. .... Black Boy abuses but at the same time identifies with the victimized animal . . ." (Allen, 1985).

The hanged cat in *Black Boy*, the crushed snake in "Big Boy Leaves Home" (Wright, 1978), the bludgeoned rat in *Native Son* (Wright, 1940)—all resemble, in some fundamental way, the humans who kill them. An animal may also be a double for the white oppressor, as is the lynch gang's bloodhound, strangled by the fugitive in "Big Boy Leaves Home." Sadly, it may simply be the victim's victim, as happens to the mule that is accidentally shot in "The Man Who Was Almost a Man" (Wright, 1978).

Wright's ability to forge a symbolic link, albeit a destructive one, between poor blacks and subjugated animals finds its most developed expression in "The Man Who Went to Chicago" (Wright, 1978). Based on the author's experiences as a laboratory-animal attendant, this story draws many parallels between experimental subjects in a hospital research facility and the four black workers who clean up after them. These men are treated by the scientists as if they are "close kin to the animals [they] tended," and a feeling of sympathy develops among them for the caged specimens. But when a fight breaks out among the attendants and their own animality comes to the surface, cages are overturned, animals are injured and killed, and the story concludes with unrelieved hopelessness.

Faulkner's use of animal doubles is much more complex. For example, Sam Fathers, the chief black figure in *Go Down, Moses* (1955), is a proud man; in his veins flows the blood of African chiefs, and the double he consciously chooses—which he reverentially terms "Grandfather"—is the huge buck that he and the boy,

Ike McCaslin, see running "free and wild and unafraid" and do not try to kill. Even so, as the boy's cousin tells him, there are moments when Sam Fathers's face wears an expression showing that he has "smelled the cage." He explains:

> *"Like an old lion or a bear in a cage," McCaslin said. "He was born in the cage and has been in it all his life; he knows nothing else. Then he smells something. It might be anything, any breeze blowing past anything and then into his nostrils. .... He hadn't smelled the cage until that minute. Then the hot sand or the brake blew into his nostrils and blew away, and all he could smell was the cage. .... That's what makes his eyes look like that ... himself his own battleground and the mausoleum of his defeat. ..."*

The image in the first line of this quotation is expanded in the next section of the novel, "The Bear," in the larger-than-life figures of the bear, Old Ben, and Lion, "the dog that's going to stop Old Ben and hold him." It is Sam Fathers who tames and trains Lion, not through gentleness but by starving and beating him. The dog submits but loses none of his ferocity and strength, fighting and killing other dogs at the slightest provocation and so gaining a mythic stature equal to Old Ben's as the day of the hunt approaches.

Lion's loyalty is to Boon Hoggabeck, the man who finally kills Old Ben. In that fatal battle, "for an instant they almost resembled a piece of statuary; the clinging dog, the bear, the man astride its back, working and probing the buried blade. It didn't collapse, crumble. It fell all of a piece, so that all three of them, man, dog, and bear, seemed to bounce once." Joined in this tableau of struggle to the death, the three figures and those around them continue to be doubled in the hunt's aftermath. Boon places the mortally wounded Lion on his own bed, beside which the dog once slept, and will let no one else touch the body. The ritually severed paw of Old Ben is placed in Lion's grave.

There are other deaths as well. Old Sam Fathers collapses after the climactic hunt and does not rally. Boon's grief for Lion is profound and violent (as shown when he grapples for the gun with McCaslin at the dog's grave); he clearly is mourning the death of part of himself. With the killing of Old Ben, the yearly hunting ritual, an event of great importance to the human characters in the story, also comes to its end.

The glorification of Old Ben and Lion places them squarely within the tradition of totemic animals in literature extending from the Norse sagas of Balder to Richard Adams's *Shardik* (1975). Death must precede apotheosis, so it is not surprising that many of the animals we have considered are heroic not so much in their deeds as in their capacity to suffer. That this should be such an important role for animals raises major questions about our stance toward them both in the arts and in life.

A work such as Holman Hunt's *The Scapegoat* (1870), its harsh colors dramatizing the ordeal of the animal driven out into the desert, clarifies the issue, reminding us of the range of ancient rituals in which human sin was transferred symbolically to an animal, which was then cast aside. In the words of Jungian psychologist Aniela Jaffé: "The animal motif is usually symbolic of one's primitive and instinctual nature. ... Even civilized men must realize the violence of their instinctual drives and their powerlessness in the face of the autonomous emotions arising from the unconscious ... " (Jaffé, 1973). It is common for these impulses to be disowned and projected upon other forces within the environment, in many cases, an animal—the quintessence of instinct. As we have seen, such symbolization may be acted out so fully as to make victims of humans and animals alike. The doubling is destructive and incomplete: The animal stands for man but only for the part of him that he fears and disavows.

Jaffé adds:

> *The boundless profusion of animal symbolism in the religion and art of all times does not merely emphasize the importance of the symbol; it shows how vital it is for men to integrate into their lives the symbol's psychic content—instinct. .... Suppressed and wounded instincts are the dangers threatening civilized man; uninhibited desires are the dangers threatening primitive man. In both cases the "animal" is alienated from its true nature and for both, the acceptance of the animal soul is the condition for wholeness and a fully lived life.....*

It is time to return to the Andokides amphora; in the light of what we have learned, its image takes on a new meaning. Heracles' twelve labors were demanded as atonement for the sin of murdering his wife and children during a fit of violent madness. The labors may also be seen as steps in his mental restoration, as stages in regaining mastery of himself. It is therefore appropriate that brute force, which first caused his undoing, should be expressly forbidden in this final challenge.

Unlike most atonement rites involving animals, this act does not involve casting-off or sacrifice of the animal but its introduction into the living world. The leopard's-head garment makes Heracles appear two-headed. Cerberus, too, is divided; the two heads pictured suggest a struggle between ferocity and acceptance. In the moment that we witness, the civilized part of each figure has won. Under the watchful eye of Athena, goddess of wisdom, Heracles has learned to make friends with the animal within himself.

References

Adams, R. 1975. *Shardik*. New York: Simon and Schuster.

Allen, M. 1985. *Animals in American Literature*, 137. Champaign: University of Illinois Press.

Burt, M. 1985. Images of Attachment: Canine Symbolism in English Art. *Animals* 6:20–4.

Burt, M. and Harding, C. 1986. Images of attachment. *National Forum 66(1)*:11–3.

Campbell, J. 1968. *The Hero with a Thousand Faces*. Princeton: Princeton University Press.

Camus, A. 1957. *L'étranger*. Paris: Gallimard.

Dostoevsky, F. [1846] 1972. *Notes from the Underground and The Double*. (J. Coulson, tr.). London: Penguin.

Dostoevsky, F. [1866] 1975. *Crime and Punishment*, ed. G. Gibian. New York: Norton.

Dostoevsky, F. [1880] 1950. *The Brothers Karamazov*, 351, 383–84, tr. C. Garnett, New York: Modern Library.

Faulkner, W. 1940. *Go Down, Moses*, 167–68, 241. New York: Modern Library.

Freud, S. 1964. Some Character Types Met with in Psychoanalytic work. In *Stories of the double*, 3, ed. A. Guerard. Philadelphia: Lippincott.

Jaffé, A. 1969. Symbolism in the Visual Arts. In *Man and his symbols*, 264–65, ed. C. Jung. New York: Doubleday.

Jung, C. 1967. *Symbols of transformation*, 2nd ed. Princeton: Princeton University Press.

Mann, T. 1936. *Stories of Three Decades*. New York: Knopf.

Rawlings, M. 1938. *The Yearling*, 427–28. New York: Charles Scribner's Sons.

Rewald, J. 1962. *Post-Impressionism from Van Gogh to Gauguin*, 398. New York: Museum of Modern Art.

Rilke, R. 1961. *Sonnets to Orpheus*, 133–34. ed. C. MacIntyre, Berkeley: University of California Press.

Steinbeck, J. 1937. *Of Mice and Men*. New York: Viking.

Steinbeck, J. 1959. *The Red Pony*. New York: Viking.

Wright, R. 1940. *Native Son*. New York: Harper and Row.

Wright, R. 1945. *Black Boy*. New York: Harper and Row.

Wright, R. 1978. *The Richard Wright Reader*. New York: Harper and Row.

Human-Animal Interactions: A Review of American Attitudes to Wild and Domestic Animals in the Twentieth Century

Stephen R. Kellert

Most of the chapters in this book deal with companion animals, such as the dog, the cat, and the horse. Wild animals are largely ignored. However, Stephen Kellert redresses this imbalance somewhat by exploring American attitudes to wildlife. He outlines the frequency of each of the prevalent attitudes in modern American society (the most common being the humanistic and moralistic on one side and the utilitarian and neutralistic on the other) and describes selected results from his exhaustive study of the distribution of these attitudes in different segments of society. He completes this analysis with the conclusions of his survey of trends in these attitudes from 1900 to 1976. In conclusion, Kellert discusses the development of children's attitudes to wild animals, which he found to differ significantly with age; for example, younger children express less interest in and concern for animals than do children in eleventh grade.

Kellert's studies have broad implications for public policy. His results have been used by many different groups, including attorneys who have used his attitude scales when selecting juries to hear cases involving wildlife. EDITOR

This paper summarizes the results of various studies, conducted over more than a decade, concerning American perceptions of animals and the natural environment. A typology of ten basic attitudes first is described to differentiate among fundamental views of animals, particularly wildlife. Next, a methodology for assessing the relative frequency of these attitudes is presented. Data follow on the distribution of the attitude types over the entire American population and among major demographic groups distinguished by age, education, income, race, residence, and so on. The occurrence of the attitude types among different age groups of children is also considered. Historical information is provided on changes in the relative prevalence of the attitudes types during the twentieth century. Finally, limited data are offered regarding preferences for various animal species and attitudes toward such controversial issues as conservation of endangered species and hunting.

The following typology of attitudes toward animals initially was based on a study of a relatively small but diverse group of persons directly involved with animals (Kellert, 1975). These attitude classifications were validated to some extent by a statistical procedure—multiple discriminant function analysis—that analyzed the answers of the classified respondents to a number of close-ended questions dealing with various people-animal issues.

This typology then was modified in the course of a small-scale national study of American attitudes toward animals (Kellert, 1976). One of the attitude types—the dominionistic—was broken into two attitudes—the utilitarian and the dominionistic. A ninth attitude type—the negativistic—subsuming hostile or indifferent feelings toward animals, theoretically was distinguished as two attitudes—the negativistic and the neutralistic. The negativistic attitude expressed active hostility or fear, while the neutralistic attitude involved passive avoidance of animals, stemming largely from indifference. Hypothetically, these ten attitudes, listed in Table 1, occur across diverse cultures and historical conditions, although they may vary considerably in content and frequency.

These attitudes primarily describe basic perceptions rather than behaviors, and they should not be identified with individuals. Rarely will all of an individual's actions be explained by just one attitude; moreover, an individual's attitudes may change over time as a result of new life experiences. Nevertheless, a certain degree of attitude stability usually is encountered in nearly every individual.

The Basic Attitude Types

The main characteristic of the *naturalistic attitude* is a strong interest in and affection for the outdoors and wildlife. Active contact with natural settings especially is valued, and, thus, the attitude is related closely to the wilderness and outdoor recreational benefits of wildlife. A sense of permanence, simplicity, and pleasure derived from unspoiled natural beauty typically are associated with this perspective. Most of all, observation of and personal involvement with wildlife are basic to the naturalistic interest in the outdoors, with animals providing the context and the meaning for active participation in natural settings.

The *ecologistic attitude*, like the naturalistic, focuses primarily on wildlife. The essential differences, however, are in the degree of personal involvement, recreational interest, and importance attached to an intellectual understanding of nature. The ecologistic emphasis is on a conceptual understanding of the interrelationships of species in the context of ecosystems. In contrast to the naturalistic concern for personal or recreational involvement with specific animals, the ecologistic interest is more in comprehending systematic

Table 1. Attitude Occurrence in American Society

Attitude	Estimated Percentage of American Population Strongly Oriented toward the Attitude	Common Behavioral Expressions	Most Related Values/Benefits
Naturalistic	10	Outdoor wildlife–related recreation—backcountry use, nature birding, and nature hunting	Outdoor recreation
Ecologistic	7	Conservation support, activism and membership, ecological study	Ecological
Humanistic	35	Pets, wildlife tourism, casual zoo visitation	Companionship, affection
Moralistic	20	Animal welfare support/membership, kindness to animals	Ethical, existence

Scientistic	1	Scientific study/hobbies, collecting	Scientific
Aesthetic	15	Nature appreciation, art, wildlife tourism	Aesthetic
Utilitarian	20	Consumption of furs, raising meat, bounties, meat hunting	Consumptive, utilitarian
Dominionistic	3	Animal spectator sports, trophy hunting	Sporting
Negativistic	2	Cruelty, overt fear behavior	Little or negative
Neutralistic	35	Avoidance-of-animal behavior	Little or negative

relationships between species and their natural environment. The approach shifts attention from individual animals to populations.

The *humanistic attitude* emphasizes feelings of strong affection and attachment to individual animals, usually pets. The animal is the recipient of feelings and emotional projections somewhat analogous to those expressed toward other people. No amount of affection, however, can compensate for intrinsic biological differences, and the animal thus is rendered something of a subhuman. Nevertheless, the humanistic attitude values animals most as basic sources of affection and companionship and even, for some people, as human substitutes. Considerable empathy for animal emotion and thought typically accompanies this perspective; as a consequence, anthropomorphic tendencies can result. The attributes and capacities of animals may be idealized, producing somewhat romanticized notions of animal innocence and virtue. The humanistic attitude toward wildlife usually involves strong affection for animals phylogenetically close to human beings, as well as for those that are large and aesthetically attractive. Species considerations relating to population dynamics generally are not present, since the concern is more for individual animals.

The primary concern of the *moralistic attitude* is the ethically appropriate human treatment of animals. Like the ecologistic attitude, it is a state of mind as much as a matter of a personal behavior or recreational involvement with animals. While the moralistic attitude is often associated with feelings of strong affection for animals, the more fundamental characteristic is a philosophical preoccupation with the nature of appropriate human contact with the non-human world. Perhaps the most basic tenet of the attitude is strong opposition to the exploitation of animals, a commitment to protect other forms of sentient life from human domination and exploitation, except in situations where survival is at stake or "higher" ends (e.g., the protection of other creatures) are served.

The *scientistic attitude's* predominant interests are the biological and physical characteristics of animals. This perspective values animals largely as objects of curiosity, study, and observation. While not necessarily accompanied by a lack of affection for animals, the scientistic attitude often fosters feelings of emotional detachment. Wild animals, for example, are interesting primarily as objects used in problem-solving, not as sources of companionship or wilderness recreation. The outlook may become reductionist, viewing life more in terms of basic processes—constituent parts—than as living creatures interrelating with other animals in the context of natural environments.

The *aesthetic attitude* primarily emphasizes the attractiveness or symbolic significance of animals: their artistic merit and beauty or their allegorical appeal as symbols.

The fundamental concern of the *utilitarian attitude* is the practical and material value of animals. A basic presumption is that animals should serve some human purpose and, thus, be sources of personal gain. This attitude is largely people oriented; animals are desirable only insofar as they produce some tangible advantage or reward. This attitude does not necessarily result in indifference or lack of affection for animals, but emotional considerations are usually subordinated to more practical concerns.

The *dominionistic attitude* centers on the satisfactions derived from the mastery and control of animals, typically, in a sporting context (rodeos, trophy hunting, bullfighting, and so forth). Animals are valued largely as challenging opponents, that furnish opportunities to demonstrate prowess, skill, strength, and, often, masculinity. The conquest of the animal signifies superiority and dominance—an expression of the human ability to confront wildness and render it submissive and orderly. The most prized animals are those considered to be fierce or cunning competitors. Major interests are in challenge, confrontation, and competition, generally involving physical domination and control. The fundamental characteristic of the *negativistic attitude* is an active dislike or fear of animals. In contrast, the *neutralistic attitude* is oriented more passively toward avoidance of animals because of indifference. Both attitudes share a sense of estrangement or emotional separation from animals. The two perspectives thus tend to encompass the view that a basic lack in affective and rational capacities distinguishes other animals from humans. Both attitudes evince little sense of kinship or affinity for animals.

The U.S. Fish and Wildlife Service of the Department of the Interior granted funds to explore more carefully the presence and strength of these perceptions among diverse social, demographic, and animal-activity groups in the 48 contiguous states and Alaska (Kellert, 1980a and c). Scales were developed to assess the relative distribution of the various attitude types. A great deal of experimentation, with many questions tested, was necessary to develop worthwhile measures. Most of the questions were dropped, for reasons such as lack of differentiating power, redundancy, or questionable relevance. This trial-and-error process was basic to construction of the scale, although many statistical techniques exist to facilitate such processes.

A total of 65 questions eventually were used to measure the attitude types; the smallest scale (the ecologistic) consisted of 4 items, while the largest (the utilitarian) included 13 questions. Whenever appropriate, the strength of the response (strongly versus slightly agree/disagree) was noted. Scale scores ranged from 0 to 11 for the ecologistic scale to 0 to 27 for the utilitarian. The independence of the scales was indicated in part by relatively small scale intercorrelations.

No useful scale was devised to measure the aesthetic

attitude. Furthermore, a neutralistic attitude scale could not be distinguished sufficiently from a negativistic scale, so only one scale was developed, incorporating elements of both negativistic and neutralistic attitudes, perhaps weighted more toward the latter.

These scales have been used to explore a variety of issues regarding American attitudes to wild animals. In this chapter, attitudes among the general population and among special subgroups, historical trends in these attitudes, and the development of children's attitudes are presented. Detailed reports of this research have appeared in other publications (Kellert, 1979; 1980a, b, and c; Kellert and Berry, 1981; Kellert and Westervelt, 1982; Kellert, 1984; Kellert, 1985a, b, c, and d; Kellert and Berry, 1987). The following discussion is a summary of these results.

Current American Attitudes to Animals

The scales used in the national survey are crude approximations of the attitude types; only in the broadest sense do they measure the true prevalence and distribution of these attitudes in the American population. Nevertheless, the relative frequency of the attitudes in the national sample was assessed by standardizing the various scale scores on a range from 0 to 1, plotting a regression line through the scale score distribution frequencies for each attitude, and using these frequency curves and regression figures to estimate the comparative popularity of the attitudes. Because particular scores on one attitude scale cannot be equated with similar scores on the other scales, this procedure indicated only roughly the relative frequency of the eight attitudes in the American population. The results of the analysis are presented in Figure 1.

These results suggest that, by a large margin, the most common attitudes toward animals in contemporary American society are the humanistic, moralistic,

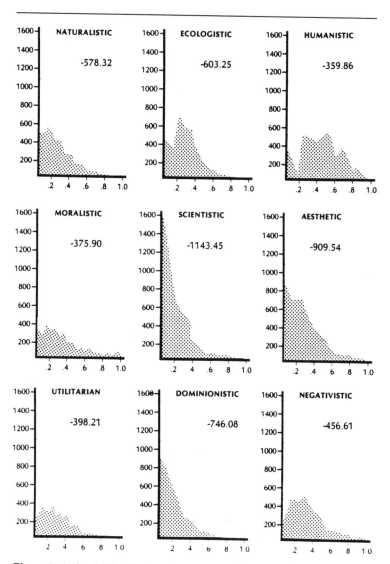

Figure 1. Attitude Distribution Curves for Nine Attitude Types in the United States Population.

utilitarian, and negativistic attitudes. In many respects, these attitudes can be subsumed under two broad and conflicting dimensional perceptions of animals. The moralistic and utilitarian attitudes clash on the subject of human exploitation of animals. The former opposes many exploitative uses of animals involving death and presumed suffering (hunting, trapping, whaling, laboratory experimentation, and so forth), while the latter endorses such utilization and other human activities that might affect animals adversely, if significant human material benefits result. In a somewhat analogous fashion, the negativistic and humanistic attitudes tend to clash, although in a more subtle fashion, on the subject of affection for animals. The former is characterized by indifference and incredulity regarding the notion of loving animals, while the latter involves intense emotional attachments to animals. The relative popularity of these four attitudes in contemporary American society may explain much of the current conflict and misunderstanding over issues involving people and animals.

The scientistic and dominionistic attitudes toward animals, according to Figure 1, are the least common among the American public. The shape of the naturalistic frequency curve suggests that this attitude is strong among a minority of Americans but relatively weak in the majority. The ecologistic scale distribution indicates that a substantial number of respondents expressed modest support for this viewpoint, but very few were oriented strongly in this direction. The estimated percentage distribution of the attitude types in the American public, the most common behavioral expressions of the attitudes, and benefits or values generally associated with each are summarized in Table 1.

The distribution of the attitudes among different demographic groups (age, sex, urban-rural residence, and income) and animal-activity groups (hunters, birders, and organization members) was also examined.

The possibility that variable differences were a function of interrelationships among certain demographic factors prompted the use of a statistical procedure—analysis of variance—to examine the combined effect of a number of demographic groups on the attitude scales. Age, sex, race, marital status, occupation, education, income, region, population of present residence, and attendance at religious services were subjected to analysis of variance.

Multiple classification analysis, a statistical technique based on analysis of variance, was then used to determine which categories of a variable contributed most to the overall significance of the factor—for example, which specific regional or educational groups are most related to the scale after all other demographic variables have been taken into account. To expedite the discussion, only the results relating to humanistic, moralistic, utilitarian, dominionistic, and negativistic attitudes are described.

The Naturalistic Attitude. A comparison of naturalistic attitude scale means among animal activity groups revealed that "nature" hunters had the highest scores, along with members of environmental protection organizations such as the Sierra Club and the Wilderness Society, and birders. The naturalistic scores of nature hunters were far higher than those of "meat" or "recreation" hunters. Antihunters, livestock raisers, and fishermen had comparatively low scores on this scale, although all animal-activity groups had higher mean scores than did the general population.

Among social demographic groups, Alaskans had the highest naturalistic scores. Other social groups with high scores included the college educated, the affluent, professionals, persons under 35, respondents from moderate-sized population areas, Pacific Coast residents, and those who rarely or never attended religious services. In contrast, the poorly educated, blacks, the elderly, low-income respondents, and persons from farm backgrounds scored substantially below the general population average on this scale. According to analysis of variance, marital status, occupation, and population of residence were not found to be related significantly to the naturalistic scale. Multiple classification analysis revealed that the most naturalistic groups were those with college and graduate-school education, Alaskan and Pacific Coast residents, respondents under 35 years of age, and persons who rarely or never attended religious service. In contrast, the least naturalistic were blacks, respondents with less than a high-school education, and persons over 56 years of age.

The Humanistic Attitude. Among animal-activity groups, members of humane and environmental protection organizations, zoo visitors, antihunters, and scientific study hobbyists scored very high on the humanistic scale. In contrast, livestock producers, nature hunters, and, surprisingly, bird-watchers had much lower scores. In light of their high scores on the naturalistic scale (with the exception of livestock producers), these latter groups apparently were far more oriented toward values relating to wildlife and outdoor recreation than toward the benefits derived from the love of animals, particularly pets.

Persons under 25 years of age, those earning between $20,000 and $35,000, females, respondents who rarely or never attended religious services, and Pacific Coast residents were the most humanistically oriented demographic groups. In contrast, farmers, persons over 76 years of age, residents of the most rural areas, and males had the lowest scores on this scale. Analysis of variance results suggested that size of town, education, marital status, and race were not significant.

The Moralistic Attitude. The demographic groups expressing the greatest moralistic concern were Pacific Coast residents, the highly educated, those engaged in clerical occupations, females, persons who rarely or never attended religious services, and respondents under 35 years of age. Least troubled by animal welfare and cruelty issues were rural residents, farmers, respondents from Alaska and the South, and males. Animal-activity groups scoring high on the moralistic scale included humane and environmental protection organization members and antihunters. Scientific study hobbyists also scored high. Recreation and meat hunters, sporting organization members, trappers, fishermen, and livestock producers scored very low on this scale.

The Utilitarian Attitude. Farmers, the elderly, blacks, and Southern respondents had the highest scores on the utilitarian scale. Persons under 35 years of age, those with graduate-school education, Alaskan respondents, single persons, and residents of areas with populations of one million or more indicated the least utilitarian interest in animals. Among animal-activity groups, livestock producers, meat hunters, and fishermen displayed an especially strong utilitarian orientation, in contrast to members of humane, wildlife protection and environmental protection organizations, and, to a somewhat lesser degree, scientific study hobbyists, backpackers, and bird-watchers.

The Dominionistic Attitude. The most dominionistically oriented animal-activity groups were trappers and all three types of hunters. Humane organization members and antihunters had the lowest scores on this scale, suggesting that differences in dominionistic perception of animals represents a basic and important distinction in the attitudes of hunters and antihunters. Zoo visitors and environmental protection organization members also had comparatively low scores on this scale.

Farmers, males, residents of Alaska and the Rocky Mountains, blacks, and those with high incomes were the most dominionistic demographic groups. Females, Pacific Coast respondents, the highly educated, clerical workers, and persons rarely or never attending religious services scored lowest on this scale. The differences on this scale between the most affluent and the most educated were in marked contrast to the similarities between these higher socioeconomic groups on the other attitude scales, suggesting that higher income and advanced education do not necessarily result in the same perceptions of animals.

The Negativistic Attitude. No animal-activity group revealed marked indifference toward or dislike of animals, although livestock producers did score only slightly above the general population mean. Interestingly, antihunters had comparatively high scores on this scale, suggesting that broad principles concerning the ethical treatment of animals were more salient considerations in their opposition to hunting than a general interest in animals. Environmental and wildlife protection organization members, scientific study hobbyists, and bird-watchers were the least negativistic. Among the demographic groups, the elderly, those of limited education, and females had the highest scores. Persons with graduate-school education, Alaska residents, respondents under 25 years of age, and those residing in areas with populations under 500 were the least negativistic in their perception of animals.

Additional Findings

Attitude profiles of selected demographic groups illustrate comparative group variations across all of the attitude dimensions. For example, respondents of limited education had considerably lower scores than the highly educated on all the attitude dimensions, with the exception of the dominionistic, the utilitarian, and the negativistic scales. These finding suggest a comparative lack of interest in and affection for animals among the least educated, except, possibly, in terms of sport and material gain. Indeed, the dramatic differences among the education groups pointed to a fundamental divergence in perceptions of animals and the natural world among various socioeconomic groups in our society.

Regional differences were also fairly significant and somewhat surprising. One of the most striking findings was the stronger interest, concern, and appreciation of wildlife among Alaskan respondents. In general, the inhabitants of western states exhibited greater appreciation and knowledge of wildlife, while those in the South manifested the least interest and concern for animals and the most utilitarian orientation.

Age and race profiles also show large variations. Differences between the very oldest and youngest respondents especially were striking on nearly every attitude dimension, particularly on the naturalistic, humanistic, and utilitarian scales. Those over 75 and under 25 years of age were similar only in their relative lack of knowledge of animals. Race results suggested a comparative lack of interest in and concern and affection for animals among black people.

Knowledge of Animals. All animal-activity groups scored significantly higher on the knowledge-of-animals scale than did the general public (See Table 2). However, bird-watchers, nature hunters, scientific study hobbyists, and members of all types of conservation-related organizations had markedly higher scores than did livestock producers, antihunters, zoo enthusiasts, sport and recreation hunters, and fishermen. Among demographic groups (Tables 2 and 3), the most knowledgeable were persons with higher education (especially graduate training), residents of Alaska and the Rocky Mountains, males, and respondents who rarely or never attended religious services. The least informed about animals—even after accounting for the interrela-

Table 2. Animal Knowledge Scale by Selected Groups:
1978 National Sample Maximum Score = 100

Animal Activity Groups		Selected Demographic Groups	
Group	Score	Group	Score
Bird-watchers	68.3	Ph.D.	67.7
Wildlf. protect.		Non-Ph.D.	
org. memb.	65.6	graduate	61.6
Nature hunters	65.3	Alaska	60.6
Scientific study	65.0	Law or medical	
Env. protect.		degree	60.4
org. memb.	64.4	College complete	56.8
Humane org.		Rocky mountain	
memb.	62.8	region	56.8
Sportsmen org.		$50,000–99,999	
memb.	62.7	income	56.7
Gen. conserv.		Professional	56.6
org. memb.	62.5	25,000–49,999 pop.	55.7
Meat hunters	57.4	General population	52.9
Fishermen	56.4	<$5,000 income	49.3
Sport/rec. hunters	56.3	Widowed	49.1
Zoo visitors	54.8	6th–8th grade	
Livestock raisers	53.9	education	47.8
Antihunters	53.9	Black	46.1
General population	52.9	75+ Years old	46.0
		<6th grade	
		education	44.4

tionships of all demographic variables—were blacks, respondents with less than a high-school education, persons over 75 and, interestingly under 25 years of age, and residents of cities with populations of one million or more.

The American public as a whole was characterized by an extremely limited knowledge of animals. For example, on four questions dealing with endangered species (see Table 4), no more than one-third of the respondents

Table 3. Animal Knowledge Scale Analysis of Variance and Multiple Classification Analysis Results Against Selected Demographic Variables

Demographic Variable	F Value
Analysis of Variance	
Age	7.67**
Population of present residence	3.09**
Region	5.93**
Education	31.83**
Occupation	0.23
Religiosity	4.75**
Income	5.31**
Marital status	3.07*
Race	30.31**
Sex	66.82**
Multiple Classification Analysis: Largest Positive and Negative Deviations After Adjusting for Independent and Covariant Variables	
Graduate education	7.73
Alaska	4.86
Rocky mountain states	2.75
College education	2.36
Male	2.18
Rarely/never attend religious services	1.96
1 Million + population	-2.07
18–25 years old	-2.30
76+ years old	-3.12
9th–11th grade education	-3.36
Less than 8th grade education	-5.10
Black	-5.50

*Significance ≤ 0.05
**Significance ≤ 0.01

Table 4. Knowledge of Endangered Species

Question or Statement	% Correct Answer	% Wrong Answer	% Don't Know
The passenger pigeon and the Carolina parakeet are now extinct.	26.2	23.1	50.6
Pesticides were a major factor in the decline of brown pelicans.	33.3	9.9	56.8
The manatee is an insect.	25.6	23.1	51.3
Timber wolves, bald eagles, and coyotes are all endangered species of animals	25.6	61.7	13.8

arrived at the correct answer—only 26% knew that the manatee in not an insect, and just 24% answered correctly the statement "timber wolves, bald eagles, and coyotes are all endangered species of animals." Regarding other knowledge questions, just 13% knew that raptors are not small rodents, and one-half of the sample answered incorrectly the statement, "spiders have ten legs." A better but still distressingly low 54% knew that veal does not come from lamb, and 57% indicated the correct answer to the question, "most insects have backbones." The knowledge questions were divided into a number of generic categories, and a comparison of mean scores revealed that the public was most knowledgeable on questions concerning animals implicated in human injury, pets, basic characteristics of animals (e.g., "all adult birds have feathers"), and domestic animals in general. On the other hand, they were least knowledgeable about endangered species, invertebrates, "taxonomic" distinctions (e.g., "Koala bears are not really bears"), and predators.

The general public was also questioned on its perceived familiarity with or awareness of eight relatively prominent wildlife issues (listed in Table 5). The three most widely recognized issues were the killing of baby seals for their fur (43% knowledgeable), the effects of pesticides such as DDT on birds (42% knowledgeable), and the use of steel leghold traps to trap wild animals (38% knowledgeable). The least familiar issues included the use of steel versus lead shot by waterfowl hunters (14% knowledgeable) and the Tennessee Valley Authority Tellico Dam/Snail Darter controversy (17% knowledgeable). The public appeared to be far more aware of emotional issues involving specific, attractive, and typically large and "higher" animals, rather than more abstract issues involving indirect impacts on wildlife due to habitat loss and dealing with "lower" animals.

Species Preference. The national sample was asked its feelings about 33 species on a seven-point like-dislike

Table 5. Awareness of Selected Wildlife Issues

Issue	% Knowledgeable	% Not Knowledgeable
Killing baby seals for their fur	43	32
Effects of pesticides such as DDT on birds	42	32
Using steel leghold traps to trap wild animals	38	37
Endangered Species Act	34	40
Killing of livestock by coyotes	23	52
Tuna-porpoise controversy	27	55
Tennessee Valley Authority	17	70
Tellico Dam/snail darter issue	17	70
Use of steel shot versus lead shot by waterfowl hunters	14	75

Note: The "knowledgeable" category combines the groups of very and moderately knowledgeable; the "not knowledgeable" category combines the groups of very little and no knowledge. The "slightly knowledgeable" category results are omitted in this comparison.

scale (see Table 6 for selected results). The most preferred were two common domestic animals—the dog and the horse—followed by three familiar and highly aesthetic creatures from two bird species and one insect order—the robin, the swan, and the butterfly. The trout—a popular and highly attractive game species—was the best-liked fish, and the most preferred wild predator was the eagle. The most favored wild mammalian species was the elephant.

Of the four least-liked animals, three were biting, stinging invertebrates—the cockroach, the mosquito, and the wasp. The third, fifth, and sixth least preferred animals—the rat, the rattlesnake, and the bat—have all been implicated in physical injury or disease inflicted

Table 6. Animal Preference Means

Most Liked Animals		Least Liked Animals		Species Preference Mean/Scores For Selected Animal Scores	
Animal	X Value[*]	Animal	X Value[*]	Animal	Mean Score[*]
Dog	1.70	Cockroach	6.45	Domestic animals	2.08
Horse	1.79	Mosquito	6.27	Attractive animals	2.38
Swan	1.97	Rat	6.26	Game animals	2.59
Robin	1.99	Wasp	5.68	Birds	2.98
Butterfly	2.04	Rattlesnake	5.66	Mammals	3.40
Trout	2.12	Bat	5.35	Amphibians, reptiles, fish	3.55
Salmon	2.26	Vulture	4.91	Predators	3.91
Eagle	2.29	Shark	4.82	Animals known to cause	
Elephant	2.63	Skunk	4.42	human property damage	4.02
Turtle	2.69	Lizard	4.13	Invertebrates	4.64
Cat	2.74	Crow	4.06	Animals known to inflict	
Ladybug	2.78	Coyote	4.02	human injury	5.08
Raccoon	2.80	Wolf	3.98	Unattractive animals	5.46
				Biting and stinging	
				invertebrates	6.13

[*]Lower score indicates greater preference.

on human beings. Relatively negative views of the coyote and wolf were noteworthy in light of the continuing controversy over predator-control programs in the United States and the considerable favorable publicity about the wolf in recent years. High standard deviation scores for the wolf, coyote, lizard, skunk, vulture, bat, shark, and cat suggested considerable variation in public opinion regarding these animals.

A qualitative assessment of the most and least preferred animals, as well as a categorical mean grouping of the 33 animals according to particular qualities (attractive, unattractive, predator, and so on; see Table 6), suggested a number of particularly important factors in public preference for different species:

1. Size (usually, the larger the animal, the more it is preferred)
2. Aesthetics
3. Intelligence (a capacity not only for reason but also for feeling and emotion)
4. Dangerousness to humans
5. Likelihood of inflicting property damage
6. Predatory tendencies
7. Phylogenetic relatedness to humans
8. Cultural and historical relationship
9. Relationship to human society (e.g., pet, domestic farm, game, pest, native wildlife, exotic wildlife)
10. Texture (generally, the more unfamiliar to humans, the less preferred)
11. Mode of locomotion (generally, the more unfamiliar to humans, the less preferred)
12. Economic value of the species

Attitudes toward Endangered Species Conservation and Hunting

Attitudes toward various animal conservation and treatment issues have also been explored (see, for example, Kellert, 1985; 1986). Results of studies of two issues are described briefly here—endangered species conservation and hunting.

Public attitudes toward the protection of endangered species were generally examined in the context of varying socioeconomic impacts including energy development, water use, forest utilization, and industrial development. The results suggest that eight factors influence critically the public's willingness to protect endangered wildlife. The first is aesthetics. The second is phylogenetic relatedness to humans—generally speaking, the closer the biological relation of the endangered animal to human beings, the greater the likelihood of public support for the species. The third factor is the reason for endangerment, with greater public sympathy typically in cases involving direct causes of endangerment (e.g., overexploitation or persecution) than in situations involving indirect impacts (e.g., habitat loss due to expanding human populations). The fourth factor is the economic value of the exploited species. The fifth concerns the numbers and types of people affected by efforts to protect the endangered animal. The cultural and historical significance of the endangered species is the sixth factor and may be involved in public sympathy for the bald eagle and trout. The seventh variable is the public's knowledge of and familiarity with the endangered animal; public support for the American crocodile may be due to this factor. Finally, the perceived humaneness of the activity threatening the species may be important; for example, the relatively slight opposition to water uses endangering an unknown fish species may stem in part from assumptions regarding the capacities of fish to experience pain.

The willingness to protect endangered wildlife varied considerably among demographic groups. Much greater willingness to protect endangered species was found among the highly educated, younger and single respondents, persons residing in areas with populations of more than one million, and residents of Alaska. Older respondents, persons with less than an eighth-grade education, farmers, residents of highly rural areas, and residents of the South expressed significantly less support for the protection of endangered species.

One very controversial issue facing the wildlife field today is the public's attitude toward hunting. Attitudes toward six different kinds of hunting were explored. The general public overwhelmingly approved of the two most pragmatically justified types of hunting—subsistence hunting as practiced by traditional native Americans and hunting exclusively for meat regardless of the identity of the hunter. On the other hand, approximately 60% opposed hunting solely for recreational or sporting purposes, whether for waterfowl or big game. Moreover, over 80% objected to the notion of hunting for trophies. Perhaps most interestingly, 64% approved of hunting for recreational purposes as long as the meat was used.

Historical Trends

The presumption of most historians is that contemporary Americans are more concerned about wildlife than ever before. But do we perhaps presume too much? Is our age truly distinctive in its degree of environmental and wildlife awareness, at least among ordinary Americans? The passage of laws often reflects the power and persuasiveness of special interest groups more than the pressing concerns of the general public—can we be certain that the many legislative changes in environmental law and protection since World War II truly indicate substantive shifts in the average person's perception of animals? It has been suggested that an analysis of the content and frequency of newspaper articles on animals over a period of time might be one approach to assessing historical trends in public attitudes (Bos et al, 1977).

Despite the tendency of newspapers to report primarily on "newsworthy" events, a number of factors recommend this approach. First, newspapers are usually oriented to local constituencies. Second, if selected judiciously, newspapers can reflect urban and rural differences, as well as regional ones. Third, by reviewing newspapers in continuous publication throughout the century, historical changes relatively undistorted by interpretive recall can be examined. Finally, because of their local and continuous publication, newspapers can reveal the experience and concerns of a large fraction of the general public. The contents of two urban—The Los Angeles *Times* and the Hartford *Courant*—and two rural newspapers—the Buffalo, Wyoming *Bulletin* and the Dawson, Georgia *News*—were sampled for the period from 1900 to 1976. On average, 2.74 animal-related articles appeared in each newspaper issue, for a total of 4,873 articles which were coded for attitude type by trained analysts (Kellert, 1982).

Somewhat unexpectedly, the numbers of animal-related articles apparently did not increase during the century (see Figure 2). The greatest number of articles in a single year was 275 in 1964, although 1921, 1927, 1930, and 1967 also had more than 200 articles. When the 75 years were distinguished according to critical historical periods, three periods averaged more than 200 articles per year: 1921–1927, 1930–1936, and 1961–1967. The periods with the fewest articles were the periods of the two World Wars—1916–1918 and 1940–1944—each averaging fewer than 115 articles per year. The analysis ended in 1976, a year that has been linked to the start of the animal rights movement; it is possible that, had the analysis continued, it would have revealed some significant changes over the past ten years.

Interestingly, the three periods with the greatest number of animal-related articles were among the four major conservation periods in the twentieth century identified by Rose (1971). The most important con-

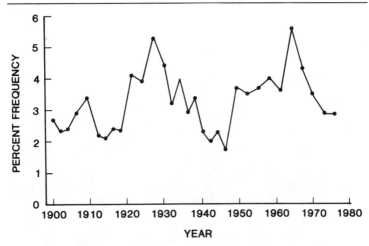

Figure 2. Frequency of Animal-Related Newspaper Articles, 1900–1976

servation-related influence of the 1920s was a techno-logical development—the automobile—that markedly stimulated public interest in wildlife and the outdoors (Trefethen 1976). For the first time, national parks and wilderness areas were readily accessible to tourists, campers, and sportsmen, sparking an unprecedented demand for recreational use of natural resources. Al-though extensive recreational interest in wildlife and the natural environment continued, the 1930s brought a new nationwide focus on the effects of grossly unwise resource use and depletion and saw the first large-scale federal attempts at wildlife and public land manage-ment. Finally, the 1960s witnessed the emergence of broad public concern for wildlife conservation, symbol-ized by the institution of Earth Day as the decade drew to a close.

The utilitarian attitude was by far the most common (48.5%) during the century, with the humanistic placing a distant second (16.1%).

During the 1970–76 period, however, the utilitarian

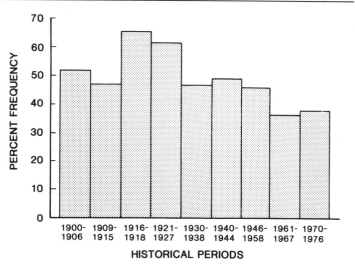

Figure 3. Frequency of Utilitarian Attitude from 1900 to 1976.

attitude accounted for only 39%, while the neutralistic increased to 27.5% of the total, mostly after World War II. The utilitarian attitude was particularly widespread during World War I, when it accounted for a remarkable two-thirds of all attitude classifications (see Figure 3). Also notable has been the substantial decline in the utilitarian attitude in recent decades; the Los Angeles *Times* showed a decline from a high of over 50% during World War I to just over 20% since 1970. In striking contrast, the utilitarian perspective decreased only slightly in the rural newspapers; in the Dawson *News*, it still accounted for nearly two-thirds of all attitude classifications from 1970 to 1976. These results are echoed in the findings on the humanitarian attitude, which increased in the Los Angeles *Times* (to 30% of all articles since World War II) and decreased substantially (to only 3% of articles during 1961 to 1976) in the rural Buffalo and Dawson newspapers.

The moralistic attitude occurs the least frequently,

present only in an average of 3% of the articles. Moreover, this attitude appeared in no more than 5% of the articles during any time period. This result was especially surprising given that the moralistic perspective was among the four most prevalent attitudes in the 1978 national survey (Kellert, 1980b). This discrepancy between historical and survey analyses reflects the difficulty of comparing findings based on widely varying methodologies. Nevertheless, the historical findings do suggest the limited importance of animal cruelty and rights considerations as newsworthy events during the century. This was particularly evident in the rural newspapers, where the moralistic attitude was practically nonexistent after World War II. It occurred far more often in the Los Angeles *Times*, however, present in 8% of the newspaper's articles since World War II. This urban/rural difference was similar to results reported in the 1978 national study (Kellert 1980b).

The frequency of various animal-related activities and the types of animals featured in the stories was also examined. Farming and livestock production were the most frequent, accounting for 25% of the total, but declined sharply after 1960. For example, farming accounted for 13% of the Los Angeles *Times* classifications at the beginning of the century but only 5% since 1961. By contrast, reports of hunting increased, and the number of game animal classifications rose from 15% before 1946 to 39% of the total since 1961. The most frequent animal category was the dog (549 articles), followed by the horse (370 articles) and the cow (360 articles). Deer (116 articles) was the most frequently mentioned wild animal (ranked 8).

Development of Attitudes Among Children

Children's perceptions of animals, particularly those of very young children, are very difficult to assess (Pomerantz, 1977). But an exploratory study of 267 Connecticut

Table 7. Relationship of Knowledge to Age, Sex, Ethnicity, and Urbanization

	P Less than	X Score	Deviation
Age	.01		
Second grade		30.50	–13.99
Fifth grade		39.64	–5.64
Eighth grade		51.18	7.45
Eleventh grade		55.11	11.57
Sex	.02		
Male		47.03	2.47
Female		41.82	–2.39
Ethnicity	.02		
White		47.40	1.17
Nonwhite		31.60	–5.72
Urban/Rural	.01		
Urban		38.00	–5.96
Small city		46.90	–1.13
Suburb		42.90	–1.13
Rural		52.30	5.23

children in the second, fifth, eighth, and eleventh grades was undertaken, evaluating their knowledge of animals and their attitudes to animals (Kellert, 1985a).

Knowledge varied widely with age, sex, ethnicity, and geographic place of residence. Eleventh-grade children had the highest scores; second graders, the lowest. Relatively high knowledge scores occurred among rural children and eighth graders. Conversely, relatively low knowledge means were characteristic of black children and children residing in large cities (see Table 7).

Age distinctions were especially impressive. Knowledge scale differences between eighth and eleventh graders, however, were substantially less divergent than between fifth and eighth graders, suggesting a decline in the effect of age on knowledge. An absence of knowledge scale differences among adults over eighteen years of age supported this hypothesis.

Ethnic differences, too, were striking, particularly the very low knowledge scores of nonwhites. Black children had the lowest knowledge scores of any demographic group with the exception of second graders. These knowledge scale differences persisted even after the possible confounding effects of other demographic variables, particularly urban/rural residence, were considered.

Urban/rural differences also were very significant, particularly when comparing children living in large cities with those residing in the most rural areas. Rural children had the second highest knowledge scale scores, while children residing in large cities had the third lowest scores.

Male/female differences were less pronounced, although still notable at .02. Significantly higher male knowledge scores occurred typically when the animal was a predator. Species preference results also revealed a more negative view of predator animals among female children.

The relative occurrence of the attitudes was assisted by examining attitude scale score frequency distributions, the slope of the regression line of the frequency distributions, and standardized attitude scale mean scores. According to these indicators, the most common attitude was the humanistic. This attitude scale had the highest mean score and the lowest slope figure (indicating a more dispersed frequency distribution) and included more children in the higher scoring ranges. Also indicative of the relative popularity of the humanistic attitude was the finding that "lovable animals" were the type most preferred, cited by 39% of the children. In general, strong emotional attachment to individual animals and a tendency toward anthropomorphism were typical among the children studied.

In the national study of adults, the humanistic attitude was also the most frequent perspective of animals, followed by the negativistic and moralistic attitudes, which were similarly popular (Table 8). The most striking difference among children and adults was in the occurrence of the naturalistic and utilitarian perspectives. The naturalistic attitude was much more common among children, while a utilitarian view of animals was far more typical of adults.

Significant age differences were observed on every scale, with the exception of the humanistic. Younger children consistently placed the needs of people over those of animals, expressing minimal concern for the rights and protection of animals. This difference was visible in highly significant results on the utilitarian, dominionistic, and moralistic scales. Younger children also expressed far less interest in animals, particularly wildlife, reflected in strong negativistic and naturalistic results. Finally, younger children were substantially less knowledgeable and informed about animals and the natural environment, as suggested by striking knowledge and ecologistic scale findings.

A more graphic indication of attitudinal changes among varying age groups is depicted in Figures 4 and 5. These figures reflect the dramatic decrease in negativistic, utilitarian, and dominionistic attitudes and the corresponding increase in the ecologistic, naturalistic, and moralistic attitudes from the second to the eleventh grades.

The results were somewhat surprising, perhaps because our society idealizes young children's perceptions of animals. The tendency is to believe young children have some natural affinity for living creatures, regarding them as little friends or kindred spirits. The results suggest otherwise; young children were the most exploitative, unfeeling, and uninformed of all children in their attitudes toward animals. Some have argued that our society creates a make-believe world for young children, and this may also involve a distorted percep-

Table 8. National Sample—18 Years and Older; Children's Sample—2nd, 5th, 8th, 11th grades by Attitude Scale Mean Scores, Slope of Scale Frequency Distribution, and Rank of Occurrence.

	Adults			Children		
	$\overline{X}$	Slope Based on Actual Response Ranges	Rank of Occurrence	X	Slope Based on Actual Response Ranges	Rank of Occurrence
Dominionistic	.14	−746.08	7	.27	−53.06	5
Ecologistic	.22	−603.25	5	.27	−61.60	5
Humanistic	.36	−359.86	1	.43	−17.09	1
Moralistic	.27	−375.90	3	.33	−37.05	4
Naturalistic	.20	−578.32	6	.35	−25.04	2
Negativistic	.28	−456.61	2	.35	−23.68	2
Scientistic	.10	−1143.45	8	.23	−76.67	8
Utilitarian	.23	−398.21	4	.30	−45.74	4

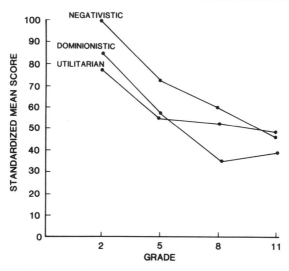

Figure 4. Attitudes that Decrease with Age

tion of young people's actual views toward animals. These results suggest that educational efforts for children six to ten years of age might focus best on the affective realm, emphasizing emotional concern and sympathy for animals.

The most profound shift between fifth and eighth grade was a major increase in factual knowledge about animals. The apparent value of emphasizing factual learning at this age is consistent with results reported by others.

Eleventh graders were far more ecologistic, moralistic, and naturalistic in their attitudes toward animals than were eighth graders. Activity results also showed that eleventh-grade children were far more interested in direct contact with and recreational enjoyment of wildlife and the out-of-doors. The most basic change at this stage, then, involved marked growth in ethical concern for animals, appreciation of wildlife, and the ability to deal with abstract concepts such as ecosystems

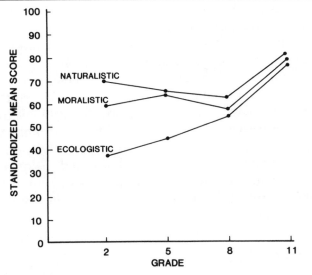

Figure 5. Attitudes that Increase with Age

and biological diversity. This period appears to offer the best opportunity for developing ethical concern for animals and an understanding of ecology.

In summary, three major transitions were identified by the results. The period from second to fifth grade was characterized most significantly by a major increase in emotional concern and affection for animals. The years between fifth and eighth grades witnessed a dramatic improvement in factual and cognitive understanding of animals. Finally, the transition from eighth to eleventh grade was marked most by a notable expansion in ethical and ecological concern for animals and the natural environment.

The attitude and knowledge scores of children who participated frequently in selected activities involving animals were also examined. Particularly surprising were the relatively low knowledge scores of children who learned about animals in school or who visited zoos; these two groups also had the highest negativistic scale scores: Apparently, these activities exerted little positive influence on children. Most zoological parks fail to go beyond superficial entertainment toward instilling greater appreciation of animals among children, while most learning about animals in school appears to be so divorced from direct experience that little increase in basic knowledge results. The positive value of direct, participatory contact between children and animals is suggested by the more encouraging activity results found among children who bird-watched, belonged to animal-related clubs, or hunted. These children were generally more appreciative, knowledgeable, and concerned about animals.

Conclusion

The four most prevalent attitudes toward animals documented in this chapter indicate a fair degree of diversity and potential conflict among the perceptions of large groups of Americans. This diversity includes a good measure of indifference and some fear of animals in a sizable segment of the public. It also involves a fair degree of support for utilizing animals for practical advantage. On the other hand, much affection for animals, particularly for pets and large, attractive wild animals, was found among many Americans. In addition, a substantial minority indicated concern for presumed maltreatment, possible suffering, and undue harm resulting from uses of animals.

While the frequency of positive feeling and concern for animals in America today is pleasing to note, the emotional rather than intellectual basis for this interest and the focus on pets and specific wildlife species pose some potential problems. As Joseph Wood Krutch once remarked, "love is not enough." An overly emotional attachment to animals may result in misguided priorities—as was suggested by the finding that the baby seal issue was of far more concern to the public than the

Endangered Species Act (Kellert, 1979). Moreover, it can lead to the over-emphasis of narrow segments of wildlife (for example, large, attractive animals) that overlooks more basic considerations of ecological relationships between wildlife and their natural habitats. Two attitudes reflecting considerable interest in wildlife and concern for natural habitats—the naturalistic and the ecologistic—were encountered far less frequently, although a naturalistic appreciation of wildlife and the outdoors was strongly present among a minority of Americans, particularly those who often catalyze change—the young, the educated, and, to some extent, western residents. Furthermore, while ecologistic concern for the natural environment was evident only among a small group of people, a fairly widespread, if somewhat unsophisticated support for the notion of conservation seemed to exist among many Americans.

These results suggest that wildlife values are undergoing a period of some confusion and transition. Creating sufficiently broad concern to support needed programs in wildlife protection and restoration may require a long and difficult campaign. Nevertheless, a bedrock of affection and concern is already present, no matter how naively expressed. The transformation of this fundamental interest to a more ecological and appreciative commitment is the challenge facing the wildlife conservation field today. It will demand considerable patience, empathy, and tolerance, as well as a willingness to be involved with many different kinds of people. The challenge is great, but so are the stakes— the future well-being of our wildlife may depend on the outcome.

References

Bos, W., Brisson, L. and Eagles, P. 1977. A Study of Attitudinal Orientations of Central Canadian Cultures towards

Wildlife. London, Ontario: University of Waterloo, Project No. 702–17.

Kellert, S. R. 1975. From Kinship to Mastery: A Study of American Attitudes toward Animals. *National Association for the Advancement of Humane Education* 2(3): 24–7.

———. 1976. Perceptions of Animals in American Society. *Transactions of the 41st North American Wildlife and Natural Resources Conference*, 533–46.

———. 1979. American Attitudes toward and Knowledge of Animals: An Update. *International Journal for the Study of Animal Problems*, 1:87–119.

———. 1980a. *Activities of the American Public Relating to Animals*, document number 024-010-00-622-2. Washington, DC: U.S. Government Printing Office.

———. 1980b. Contemporary Values of Wildlife in American Society. In *Wildlife Values*, 31–60, ed. W. W. Shaw and E. H. Zube, Center for Assessment of Noncommodity Natural Resources, Rocky Mountain Forestry and Range Experimental Station, U.S. Forestry Service, Fort Collins, CO.

———. 1980c. *Public Attitudes toward Critical Wildlife and Natural Habitat Issues*, document number 024-010-00-623-4. Washington, DC: U.S. Government Printing Office.

———, and Berry, J. 1981. *Knowledge, Affection and Basic Attitudes toward Animals in American Society*, document number 024-010-00-625-1. Washington, DC: U.S. Government Printing Office.

———, and Westervelt, M. 1982. *20th Century Trends in American Perceptions and Uses of Animals*, document number 024-010-006-218. Washington, DC: U.S. Government Printing Office.

———. 1983. Affective, Evaluative and Cognitive Perceptions of Animals. In *Behavior and the Natural Environment*, 241–67, ed. I. Altman and J. F. Wohlwill. New York: Plenum Press.

———, and Westervelt, M. 1983. Children's Attitudes, Knowledge and Behaviors toward Animals, document number 024-010-00-641-2. Washington, DC: U.S. Government Printing Office.

———. 1984. Urban American Perceptions and Uses of Animals and the Natural Environment. *Urban Ecology* 8:209–28.

———. 1985a. Attitudes toward Animals: Age-Related Development among Children. *Journal of Environmental Education* 16(3):29–39.

————. 1985b. Historical Trends in Perceptions and Uses of Animals in 20th Century America. *Environmental Review 9(1)*:19–33.

————. 1985c. Public Perceptions of Predators, Particularly the Wolf and Coyote. *Biological Conservation 31*:167–89.

————. 1985d. Socioeconomic Factors in Endangered Species Management. *Journal of Wildlife Management 49*:528–36.

————, and Berry, J. 1987. Attitudes, Knowledge and Behavior toward Wildlife as Affected by Gender. *Bulletin of the Wildlife Society*. In press.

Pomerantz, G. A. 1977. Young People's Attitudes toward Wildlife. Lansing, MI: Michigan Department of Natural Resources, Wildlife Division Report 2781.

Rose, H. M. 1971. Conservation in the United States. In *Conservation of Natural Resources*, ed. G. H. Smith. New York: John Wiley & Sons.

Trefethen, J. B. 1976. *The American Landscape: 1776–1976; Two Centuries of Change*. Washington, DC: Wildlife Management Institute.

Summation: People, Animals, and the Environment

Bruce Fogle

Every collected work requires a suitable concluding chapter. Sometimes, this chapter is written by the editor, but, in this case, one of the pioneers of human-animal–bond studies has been pressed into service. Bruce Fogle leads the reader through a well-written examination of why modern Western society has been so slow to study the role that pets play. He speculates that people have, until recently, placed animals very low on the scale of considerability because of the cultural influence of Aristotle's chain of being (which puts humans at the top) and Cartesian mechanism (which denies animals any conscious experience). As a result of these prevailing intellectual biases, the notion that an animal's companionship might be important has received very little serious consideration. However, the chasm once thought to exist between humans and animals is being narrowed due to a resurgence in cognitive ethology and the growth of sociobiology. Cognitive ethology has shown that animals are more than unfeeling, stimulus-response machines, while sociobiology emphasizes that we may not be as far removed from our animal roots as we might think. Fogle stresses the need for more study of human-animal interactions and concludes by asking whether people are more likely to be good people because they are caring for and nurturing animals or whether they are more likely to care for animals because they are already good people. Human-animal bond research is full of such cause-and-effect conundrums. EDITOR

When I speak to veterinarians or to veterinary students about the relationship between people and their companion animals, I catch myself telling them that this is a new field of study. Describing this subject as a new area of interest is a neat way to avoid the risk of their applying what I have said to any circumstances other than the specific situations that I discuss. My admonition carries with it the unspoken proviso that everything I say may be like an ice castle, massive, sturdy, and unchallengeable when built but lost as vapor in the searing heat of scrutinizing sunlight.

This was certainly the case at the first international meeting on "The Human–Companion Animal Bond" in London in 1980, and it continued to be true throughout the further major meetings in Philadelphia, Vienna, and elsewhere. But the situation changed in Boston in 1986. At that conference, we found that we now possess irrefutable facts and that our new challenge is how to interpret them.

There is, of course, something new in discussing people's relationships with companion animals—dogs, cats, horses, birds—rather than those with agricultural species or, more grandly, animals in nature. Ethology has looked upon domesticated animals as the ugly cousins of their brethren living in a more "natural" environment where the vagaries of weather and natural selection dictate who survives and multiplies. Because domestic animals do not have to forage for their food or compete for sexual supremacy, until the last decade their behavior was of little interest to anyone other than their keepers. Similarly, as James Serpell has so lucidly described, anthropologists have been interested in the role of pets among the B'Mbuti Pygmies in Zaire or the Cree Indians in northern Alberta but certainly not among the station-wagon tribe of the suburbs of Chicago or London. To my knowledge, Constance Perin (1981) in Boston, was the first anthropologist to look seriously at the role of companion animals in modern American culture.

The reasons for our late arrival at an interest in the roles that companion animals play in our lives are manifold. First, it is only with twentieth-century affluence that we have reached a position in the Western world where we can live commensally with other animals, not looking upon them as a possible source of nourishment. But there is another more pervasive reason why we have not, until relatively recently, shown an interest in the relationship we have with other animals and with our environment: our cultural disassociation from the rest of nature.

Some of the world's great religions and many so-called primitive cultures have looked upon the world as a unified whole—a creation where all was interrelated and interdependent. The primitive cultures depended on animals for survival, and their myths and stories, their teachings, and their religions, interpreted animal behavior in terms of its specific benefit to them. The Kitsumkalum Indians in British Columbia, for example, considered the annual salmon migration to be the yearly fulfillment of a contract between the hunter and the hunted, a contract designed primarily to benefit the Kitsumkalum Indians. Further inland, the Nootka tribe developed what appears to have been a genuine kinship with animals, specifically, with the wolf. The wolf was admired, emulated, and adopted as their totem, appreciated for its courage in defending the pack, its keen sense of its own territory, its stoicism, and its intense loyalty to its family. The Nootka shared their environment with the wolf, and, in many ways, it is not facetious to say that their respect for the timber wolf is echoed today in the relationship that many people have with the wolf descendants that occupy so many Western homes.

For thousands of years, our learned behavior and attitudes toward animals have been shaped by our own cultural totem, the Bible, a record of the history and the myths of a small tribe of people from the eastern shores of the Mediterranean. Our great totem is explicit about the relationship that it decrees between people, animals, and the environment. After the flood, God blessed Noah and said to him, "And the fear of you, and the dread of you, shall be upon every beast of the earth, and upon every fowl of the air, upon all that moveth upon the earth, and upon all the fishes of the sea; into your hands are they delivered." The Old Testament does envisage for the Israelites the mythical time when the wolf will dwell with the lamb and the lion will eat straw like the ox, and there are, too, prohibitions against wanton cruelty to animals. But the basis for our cultural tradition is firmly set in Genesis, where man is described as supreme and unique.

In placing such importance on the human soul, Christian thought further distanced the human from all other animals. According to fundamental Christian belief, only the human is destined for life after death. This cultural tradition still holds in the post-secular era. In a survey I conducted with a colleague on pet loss and human emotion, 43% of the 169 British veterinarians who participated stated that they believe in an afterlife for humans, while only 18% believe in an afterlife for non-human animals.

Our cultural base is also influenced by the school of the Greek philosopher, Aristotle. Anticipating Charles Darwin by a few thousand years, Aristotle attempted to illustrate in his "ladder of life" the relationships between different living creatures. He did not deny that man is an animal—indeed, he defines man as a "rational animal"—but, "since nature makes nothing purposeless or in vain, it is undeniably true that she has made all animals for the sake of man." This explanation is still central to the Western attitude to animals and the environment. Not until the seventeenth and eighteenth centuries did people such as John Ray and Carl von Linne begin cataloging nature. Gilbert White wrote the first English book on the interrelations of the environment, *The Natural History of Selbourne*. But even to these first great observers of nature, animals were but immutable acts of Creation, put on earth to serve us. Also in the seventeenth century, René Descartes inaugurated the era of modern philosophy. He believed that everything made of matter was governed by mechanical principles. The only difference between a clock and a cow, said Descartes, was that man was the mechanic of the former and God, in his infinite complexity, was the mechanic of the latter. However, according to Descartes, consciousness does not consist of matter, and consciousness is what separates humans from other animals and the rest of the environment. Consciousness survives in our immortal souls, and, of all material things, only humans have souls; all other animals are machines, automata, without the ability to feel pleasure or pain, without consciousness.

Descartes's nullification of the environment was incorporated into traditional nineteenth-century medicine and still reverberates strongly in both medicine and the social sciences. For a great many scientists, to prove that the companionship of animals can be beneficial to some people, experiments must be undertaken where the subject is isolated from all extraneous influences. What happens to subjects' heart rates after they stare at a blank wall for ten minutes in an empty room and then stroke their dogs? What changes, if any, occur to subjects' blood pressure when they stroke, but do not speak to or look at their pets?

But how can something as complex as a person's interaction with the environment be studied in the laboratory? How do you create ethological situations that will yield data subtle enough to explain how people interact with plants or animals?

Sociobiologists try to explain human behavior wholly

in terms of biological heritage. Social psychologists and cultural anthropologists try to relate all human behavior to the influences of our culture, our traditions, and fellow humans. All are correct in their limited way but isolated by the narrowness of their vision.

Meetings such as the 1986 Boston conference on "People, Animals, and the Environment" have the unique advantage of drawing scientists, academics, and professionals from many different spheres. On the first day of this conference, a message from Konrad Lorenz was read reminding us that, as the "most powerful element in creation," we have a responsibility for all animals and the environment. Other speakers at this conference included a clinical psychologist, who discussed how dog behavior can be either influenced by or indicative of the owner's personality and attitudes; a dean of a nursing college who described research indicating that pet dogs may ease the grieving process for some widows; and a social worker, who recounted her work suggesting that strangers are more at ease and likely to talk to or acknowledge people confined to wheelchairs if the physically disabled person is accompanied by a dog.

The Boston meeting reinforced the self-evident fact that we are social animals. We are, as Aristotle said, rational animals, but we are also products of our biology. As an observer of animal behavior and, more specifically, as an observer of the relationships between people and companion animals, I believe that it is obvious that we have to leave the laboratory and enter the real world to understand these relationships. I come home from work; my new dog hears the key in the door and comes to greet me. The next day, I come home from work, and my new dog is waiting to greet me at the door. The following day, I come home from work with a dog chew to give my dog, who is waiting at the door to greet me. My dog is a social influence on me, and I am an influence on her. The relationship cannot be studied in the laboratory.

I was gratified, shortly after I started to practice veterinary medicine, to read what the psychologist, Elliot Aronson, called "Aronson's First Law": "People who do crazy things are not necessarily crazy." I was pleased because I was seeing in my practice what I thought were a few crazy things—people who seemed to worry as much about their pets as they did about their children, their own genes! Aronson's first law says that most of us are susceptible to influence under the right social or psychological conditions. To that can be added the fact that we are predisposed biologically to be susceptible to these influences.

We are the most intensely social and gregarious of all living mammals. We crave companionship and suffer when it is not forthcoming. James Lynch, in his book, *The Broken Heart*, said:

> *Those individuals who lack the comfort of another human being may very well lack one of nature's most powerful antidotes to stress. Individuals who live alone—widows and widowers, divorcees, and single people—may be particularly vulnerable to stress and anxiety. The presence of a friend or a companion may not only help to suppress fear and physical pain, but it may also reduce the 'wear and tear' on the heart that occurs under stress and chronic anxiety. The increases in cardiac death rates for those who live alone may be due, in part, to the fact that these individuals continuously lack the tranquillizing influence of companionship during life's stresses. Tranquillizing drugs may not be able to fill the void, and in the final analysis they may not be anywhere near as effective as the calming capacity of friendship.*

James Lynch was writing about human companionship and human friendship, but it would be difficult to deny today that companion animals serve this role for a significant number of people. The opposite is also true. The restorative and protective effects of companionship are not a human monopoly; other social species also benefit physiologically from human contact; a human, for example, is an eminently good substitute for another dog. But we should question whether we are as good a substitute for the individuals of other species that we keep as pets. Can we, for example, fulfill the social needs of singly caged birds?

And this leads to a conundrum. Michael McCulloch, honored in this volume, was a good man, and his untimely death saddened those of us who knew him and enjoyed his company. A number of those active in the human-animal–bond field might argue that one reason why Michael was such a good man was that he was born into a family that kept pets, was raised with pets, and kept a menagerie of pets in his family. I heard one conference speaker describe his own research and conclude that students raised in families with dogs or cats are more sociable and have greater tolerance and a greater feeling of personal worth. Now that may very well be true. As a practicing veterinarian, I know that the pet owners I see are sociable and responsible people. But do they have these attributes because they were raised with and now keep domestic pets? Is the pet the cause of these attributes or is the pet merely a manifestation of the sense of well-being that so many pet owners possess? Was Michael McCulloch a good man because he grew up with pets or was he a good man because he was born and raised in a nurturing family with the type of parents who choose to keep pets?

That, to me, is the seminal question that now emerges to confront us. We know that pets improve the quality of life of people in institutions. But how can we prove cause and effect? Is there a sex difference in attitudes toward animals? Are certain people either biologically or socially primed to be more contented and nurturing and good? If so, are pets a cause of this behavior or are they a manifestation of it? Cause or effect—that is the conundrum, and its resolution is the major task that we should pursue in the future.

References

Lynch, J. J. 1977. *The Broken Heart: The Medical Consequences of Loneliness.* New York: Basic Books.
Perin, C. 1981. "Dogs as symbols in human development." In *Interrelations Between People and Pets,* 68–88, ed. B. Fogle. Springfield, IL: Charles C. Thomas.

Annotated Bibliography

Compiled by Andrew N. Rowan

This collection of books and articles is not intended to be exhaustive and, indeed, does not contain many of the items that deserve to have been included. The selection reflects the editor's own knowledge and bias towards multidisciplinary scholarship. Anyone who takes the trouble to read these publications will gain a clear perception of both the difficulties of research on human-animal interactions and the fascination of the field.

The bulk of the bibliography consists of monographs or of collected works, including the proceedings of the various international and national conferences. These include the 1974 conference in London (Anderson, 1975), the 1980 conference in London (Fogle, 1981), the 1981 conference in Philadelphia (Katcher and Beck, 1983), the 1983 conferences in Minnesota and California (Anderson et al., 1984), and the 1983 conference in Vienna (IEMT, 1985). These collected works cover much of the field and provide outlines of nearly all relevant research and theoretical writing. However, a few articles and issues either are not covered in these books or are sufficiently important to warrant inclusion in any brief bibliography. Some provide a solid and accessible review of the literature. Occasionally, it will be easier to find an article in a widely circulated periodical than to locate a copy of some of the books in this list.

Finally, the lack of a good review of children's atti-

tudes to animals is one of the biggest gaps in this literature. A few articles have been identified that cover some
aspects of this topic.

Books (Monographs)

Allen, M. 1985. *Animals in American Literature*. Champaign:
University of Illinois Press. An introduction to the extraor-
dinary menagerie found in literature of all types, especially
children's books.
Beck, A. M. and Katcher, A. H. 1983. *Between Pets and People*.
New York: Putnam. Written for the general reader with
many interesting anecdotes and stories.
Bustad, L. 1980. *Animals, Aging, and the Aged*. Minneapolis:
University of Minnesota Press. The role that animals play
in the lives of the elderly.
Clutton-Brock, J. 1981. *Domesticated Animals from Early Times*.
Austin: University of Texas Press. An easy-to-read and
well-produced introduction to the topic of domestication.
Evans, E. P. [1906] 1987. *The Criminal Prosecution and Capital
Punishment of Animals*. London: Faber and Faber. A fasci-
nating insight into the Middle Ages, when locusts and pigs
were brought to trial.
Fogle, B. 1983. *Pets and their People*. New York: Penguin. Writ-
ten for a wide audience, including anecdotes from the
author's London veterinary practice.
Harris, M. 1985. *Good to Eat: Riddles of Food and Culture*. New
York: Simon and Schuster. A materialistic and utilitarian
analysis of food habits and taboos around the world.
Lawrence, E. A. 1982. *Rodeo: An Anthropologist Looks at the
Wild and the Tame*. Knoxville: University of Tennessee
Press. Rodeo epitomizes western society in the U.S.A. An
anthropologist teases apart its cultural symbolism and discusses the conflicting values relating to the wild and the
tame.
Lawrence, E. A. 1985. *Hoofbeats and Society: Studies of Human-
Horse Interactions*. Bloomington: Indiana University Press.
The author brings to bear her own scholarship and long-
standing love of horses to analyze a variety of facets of
human-horse interactions.

Levinson, B. 1969. *Pet-oriented Child Psychotherapy.* Springfield, IL: Charles C. Thomas. The first book on pet-facilitated therapy.

Levinson, B. 1972. *Pets and Human Development.* Springfield, IL: Charles C. Thomas. The sequel to Levinson's 1969 landmark publication.

Midgley, M. 1984. *Animals and Why They Matter?* Athens: The University of Georgia Press. An excellent, readable analysis of a variety of ethical issues involving the moral status of animals.

Ritvo, H. 1987. *The Animal Estate: The English and Other Creatures in the Victorian Age.* Cambridge: Harvard University Press. The author shows how attitudes to animals changed during the nineteenth century and how selective breeding and the "dog fancy" influenced those changes.

Serpell, J. 1986. *In the Company of Animals.* Oxford, England: Basical Blackwell. An analysis of the paradox between pet-keeping and the killing and consumption of animals. The author explores pet-keeping across a variety of cultures.

Thomas, K. 1983. *Man and the Natural World: Changing Attitudes in England 1500–1800.* London: Allen Lane. A masterful analysis of changing British attitudes to animals and nature up to the end of the eighteenth century.

Tuan, Y. 1984. *Dominance and Affection: The Making of Pets.* New Haven: Yale University Press. An examination of the narrow line that divides "pets" and "companion life forms," be they plants, fish, or dogs.

Zeuner, F. E. 1963. *A History of Domesticated Animals.* New York: Harper & Row. The standard text on animal domestication.

Books (Collections)

Anderson, R. K., Hart, B. L., and Hart, L. A., eds. 1984. *The Pet Connection: Its Influence on Our Health and Quality of Life.* St. Paul, MN: Grove Publishing. The proceedings of two conferences held in Minnesota and California in 1983. The topics include the history of the human-animal bond, the effects of pet death, children and animals, animals in institutions, and several other articles on the human-animal bond.

Anderson, R. S., ed. 1975. *Pet Animals and Society.* London: Bailliere Tindall. The proceedings of one of the earliest conferences on human-animal interactions, held in London in 1974.

Arkow, P., ed. 1984. *Pets and Animals in the Helping Professions: Dynamic Relationships in Practice.* Alameda, CA: Latham Foundation. A multiauthored work providing some theory but also much practical advice on establishing pet-therapy and pet-visitation programs.

Cusack, O., and Smith, E., eds. 1984. *Pets and the Elderly: The Therapeutic Bond.* New York: Haworth Press. A collection of papers reporting and discussing research on the therapeutic value of animals for the elderly.

Fogle, B., ed. 1981. *Interrelations Between People and Pets.* Springfield, IL: Charles C. Thomas. The proceedings of the 1980 London conference on the human-animal bond.

IEMT, ed. 1985. *The Human-Pet Relationship.* Vienna: Institute for Interdisciplinary Research on the Human-Pet Relationship. The proceedings of the 1983 Vienna conference on the human-animal bond, held to honor Konrad Lorenz.

Katcher, A., and Beck, A., eds. 1983. *New Perspectives on Our Lives with Companion Animals.* Philadelphia: University of Pennsylvania Press. The proceedings of the first major American conference on the human-animal bond, held in Philadelphia in 1981. Over 60 authors are included in the volume.

Quackenbush, J., and Voith, V., eds. 1985. *Veterinary Clinics of North America: Human-Companion Animal Bond.* Vol. 15, 283–469. A selection of invited contributions on the whole range of human-animal–bond topics of interest to veterinarians.

Regan, T., ed. 1986. *Animal Sacrifices: Religious Perspectives on the Use of Animals in Science.* Philadelphia: Temple University Press. The proceedings of an international conference held in London. Scholars of all major Western and Eastern religious traditions discuss the place and the treatment of animals as elucidated by major texts and teachings.

Articles

Beck, A. M., and Katcher, A. H. 1984. A New Look at Pet Facilitated Therapy. *Journal of the American Veterinary Medication Association, 184*:414–21. A critical review of the extent to which pets have been demonstrated to have a beneficial effect. While anecdotal reports remain supportive, the larger studies nearly all suffer from methodological flaws or report results of only marginal significance.

Bryant, C. D. 1979. The Zoological Connection: Animal Related Human Behavior. *Social Forces, 58*:399–421. A sociologist examines human-animal interactions. One of the very few papers on this topic by a card-carrying sociologist, who offers some interesting insights.

Burghardt, G. M., and Herzog, H. A. 1980. Beyond Conspecifics: Is Brer Rabbit Our Brother? *Bioscience, 30*:763–68. Two psychologists examine why some animals elicit our sympathies while others evoke revulsion. An important and well-written analysis.

Felthous, A. R., and Kellert, S. R. 1987. Childhood Cruelty to Animals and Later Aggression against People: A Review. *American Journal of Psychiatry, 144*:710–17. A seminal review. The authors found ten studies arguing that there was no correlation and four arguing that there was. Methodological flaws were identified in the group of ten, and the review concludes that there probably is an association between cruelty to animals and later aggression against people.

Gould, S. J. 1979. Mickey Mouse Meets Konrad Lorenz. *Natural History, 88*:30–36. Another beautifully written essay by Gould, discussing neoteny and the releaser stimuli that promote adult nurturance of infants. Walt Disney's artists modified Mickey to make him "cuter." Must be read.

Halverson, J. 1976. Animal Categories and Terms of Abuse. *Man, 11*:505–16. A stinging critique of Leach's article (see below). Must be read with Leach.

Kellert, S. R. 1985. Attitudes toward Animals: Age-Related Development among Children. *Journal of Environmental Education, 16(3)*:29–39. An analysis of the development of children's attitudes to wildlife.

Kidd, A. H., and Kidd, R. M. 1985. Children's Attitudes toward Their Pets. *Psychology Reports*, 57:15–31. A paper on children and animals.

Kidd, A. H., and Kidd, R. M. 1987. Seeking a Theory of the Human/Companion Animal Bond. *Anthrozoös*, 1:140–57. An attempt to develop a comprehensive theory for the human-animal bond, with comments from five scholars in the field.

Lawrence, E. A. 1986. Neoteny in American Perception of Animals. *Journal of Psychoanalytic Anthropology*, 9:41–54. The author shows how American culture tends to favor childlike qualities in companion animals and animal symbols.

Leach, E. R. 1964. Anthropological Aspects of Language: Animal Categories and Verbal Abuse. In *New Directions in the Study of Language*, 23–64, ed. E. Lennenberg. Cambridge, MA: MIT Press. The author identifies three categories of verbal abuse—obscenities, profanities, and animal terms—and asks why animal terms should have the same linguistic potency as terms referring to God and sex. This paper must be read in conjunction with Halverson (see above).

Menninger, K. A. 1951. Totemic Aspects of Contemporary Attitudes toward Animals. In *Psychoanalysis and Culture*, 42–74, ed. G. B. Wilbur and W. Muensterberg. New York: International Universities Press. A psychoanalyst looks at some of the psychopathologies of human-animal interactions.

Schowalter, J. E. 1983. The Use and Abuse of Pets. *Journal of the American Academy of Child Psychiatry*, 22:68–72. The author discusses children's attitudes to pets.

Selby, L. A., and Rhoades, J. D. 1981. Attitudes of the Public towards Dogs and Cats as Companion Animals. *Journal of Small Animal Practice*, 22:129–37. The authors report on the reasons why people keep dogs and cats and the proportions of people in America who like dogs and cats.

Shell, M. 1986. The Family Pet. *Representations*, 15:121–53. An interesting paper on the concept of "pethood"—a human social institution in which animals happen to play a part.

Wilson, C. C., and Netting, F. E. 1983. Companion Animals and the Elderly: A State-of-the-art Summary. *Journal of the American Veterinary Medical Association*, 183:1425–29. A solid review of the literature on pets and the elderly.